An Introduction To

Tissue-Biomaterial Interactions

For my friend Tung –
thank you for the opportunity
to share classes with you.
Please keep in touch.

Kay C Dee
5/5/04

An Introduction To

Tissue-Biomaterial Interactions

Kay C Dee, Ph.D.
Tulane University
Department of Biomedical Engineering
New Orleans, Louisiana

David A. Puleo, Ph.D.
University of Kentucky
Center for Biomedical Engineering
Lexington, Kentucky

Rena Bizios, Ph.D.
Rensselaer Polytechnic Institute
Department of Biomedical Engineering
Troy, New York

A John Wiley & Sons, Inc., Publication

Published by John Wiley & Sons, Inc., Hoboken, New Jersey.
Published simultaneously in Canada.

Library of Congress Cataloging-in-Publication Data:

Bizios, Rena.
An introduction to tissue-biomaterial interactions / Rena Bizios, Kay C. Dee, David A. Puleo.
p. cm.
ISBN 0-471-25394-4
1. Biomedical materials. 2. Biocompatibility. I. Dee, Kay C. II. Puleo, David A. III. Title.
R857 .M3 B595 2003
610′.28—dc21 2002005591

Printed in the United States of America

10 9 8 7 6 5 4 3 2 1

To students
—of the past, present and future—
who have shared our fascination with the tissues, cells, and molecules
of the human body.

Contents

Preface

Undergraduate biomedical engineering programs and curricula are being rapidly developed across the United States. We perceive a correspondingly increasing need for biomedical engineering textbooks specifically designed for undergraduate readers. Many educators have come to appreciate that physiology and biology are not narrow, specialized applications to be "tacked onto" an engineering curriculum, but are instead rich subjects that can naturally elicit and benefit from the kinds of creative problem-solving and quantitative analyses that are hallmarks of engineering. However, integrating life sciences within the structured and rigorous framework of fundamental knowledge required for an undergraduate engineering degree—especially early in the undergraduate curriculum, before the senior year or a "capstone" course—is still a challenge for educators.

We believe that providing undergraduates with an opportunity to learn biomedically oriented material early in their academic careers establishes a vivid framework of "vocational relevance" (i.e., showing students that the basic science and engineering skills they are learning are crucially important to biomedical science) that is often otherwise lacking. We also believe that helping students see connections between seemingly disparate course materials—fluid mechanics and cell biology, for example, connected by understanding both *how* red blood cells flow through capillaries and *why* altered flow characteristics may be of clinical importance—can help students develop creative thinking and problem-solving skills. The fields of biomaterials and cell/tissue engineering present excellent opportunities to integrate life sciences and engineering, by capitalizing on the inherently interdisciplinary interface between cells/tissues and biomaterials (whether man-made or biologically-derived).

In accordance with the principles and needs described above, we have designed this textbook, which focuses on the wound healing process and interactions between the human body and implanted biomaterials or devices. This book is short, accessible, and we hope affordable, because it is intended for use in a one-semester, undergraduate-level course by students who have completed some science/engineering course work (introduction to materials, statics, and perhaps mechanics of materials or a fundamental fluid mechanics course) with minimal chemistry or biology (general chemistry and one semester of cell biology would be sufficient). Although this book is particularly appropriate for use in biomedical engineering curricula, especially those with a focus on cell or tissue engineering, we intended it to be an accessible and useful resource for other programs of study as well (premedical students with little or no engineering background, for example, or first-year engineering graduate students with little or no biomedical background). Therefore, skills in advanced mathematics and computational modeling are not required for use of this text.

This book is not intended to be a comprehensive or advanced reference (other excellent reference texts are currently available). This book is instead intended to be a resource for exploration and discovery by undergraduate students. We would encourage instructors to supplement the core material presented in this text—as we ourselves do when we teach—with relevant quantitative analyses/models, current ideas and results from recent peer-reviewed research articles, information from current events and the popular media, and advanced course work.

Acknowledgments

We thank our colleagues and students for their support and encouragement and especially thank the following: Amanda Filanowski (Tulane University) for drafting the intimal hyperplasia and osseointegration sections of this text; Luna Han and Colette Bean (John Wiley & Sons, Inc.) for constructive suggestions, understanding, and gentle insistence regarding deadlines; Glen A. Livesay, Ph.D. (Tulane University) for advice on schematic illustrations and for allowing his "Bezier Man" figure to appear in some of the illustrations; John David Larkin Nolen, M.D., Ph.D., MSPH (Emory University) and Joel Bumgardner, Ph.D. (Mississippi State University) for reviews and helpful comments; and Sue, Nick, and Angela Puleo for their patience, support, and love. Most importantly, we thank the students of Rensselaer Polytechnic Institute Course 31.4240/BMED-4240 (and earlier versions under various titles), of Tulane University Course BMEN 340/740, and of University of Kentucky Course BME 662 for their patience and good will as the concepts and material in this book were developed and refined.

A portion of the material included in this book is based on work supported by the National Science Foundation under Grant No. 9983931. Any opinions, findings and conclusions or recommendations expressed in this material are those of the author(s) and do not necessarily reflect the views of the National Science Foundation (NSF).

Introduction

The interactions of tissues and body fluids with biomaterials or medical devices is an area of crucial importance to all kinds of medical technologies. For example, electrical sensors or drug delivery patches applied externally, on the skin, must be designed to function optimally without causing skin irritation or hypersensitivity responses. Many kinds of reconstructive medical implants (hip replacements, for example, or dental implants) need to integrate with surrounding tissues to restore adequate function, without releasing harmful chemical products or significantly modifying the local electrical and mechanical environment. Pacemaker leads, arterial grafts, and dialysis machines are all further examples of devices that involve man-made materials interacting with tissues and/or body fluids like blood. This textbook is intended to help students discover how many of the macroscopic, tissue-level events (bone resorption or growth, blood clotting, fibrous tissue encapsulation, etc.) that often determine the success or failure of the medical devices listed above are, ultimately, derived from cellular and molecular level interactions with the tissue-implant interface.

A crucial concept to understand about the tissue-biomaterial interface is that a lot of things happen there! The environment inside the body is chemically, electrically, and mechanically active, and the interface between an implanted biomaterial and the body is the location of a variety of dynamic biochemical processes and reactions. For example, Figure 0.1 shows some of the atomic and molecular level events that happen when a metallic implant is placed in the body. Oxygen diffuses from the surface oxide into the bulk metal, and metal ions can diffuse from the bulk into the surface oxide as well. Biological ions can also be incorporated into the surface oxide. Interactions of biological molecules (proteins, enzymes, etc.) with the implant surface can cause transient or permanent changes in the conformation—and thus the function—of these molecules. Chapters 1 through 3 of this book provide more detail about the molecular level events that happen at the tissue-implant interface, whereas Chapters 4 through 7 explore selected biological and physiological consequences of these events. Specifically, because virtually all implantation procedures create wounds (i.e., some form of surgery is required to implant a device), tissue-implant interactions will be largely influenced by the body's wound healing response. From the first contact of biological molecules with an implant surface to the final tissue remodel-

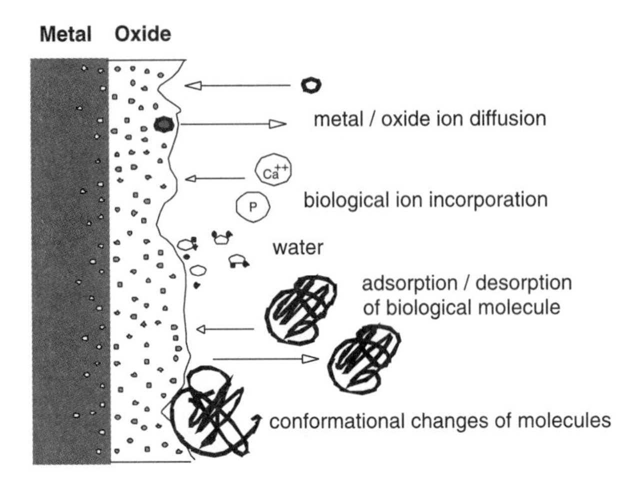

Figure 0.1. Molecular level events at the surface of a metal implant. On a macroscopic level, the surface of a metal implant may appear to be smooth, uniform, and inert. On the microscopic level, such a surface probably varies in chemical composition and topology and is the location of a number of dynamic molecule-surface interactions. A number of these molecule-surface interactions can have far-reaching physiological effects (initiating the process of coagulation, for example) that are relevant to the wound healing process and to the long-term viability of the implant.

ing around an implant, understanding the temporal progression of wound healing (Fig. 0.2) is a necessary part of understanding tissue-implant interactions.

Although the interface between a man-made, synthetic biomaterial and the body is complex enough (Fig. 0.3), tissue-engineered products that incorporate living cells within a biomaterial matrix make things even more complicated. These tissue-engineered products have three distinct interfaces to consider: between the body and the biomaterial, between the body and the living cells, and between the cells and the biomaterial (Fig. 0.4). Each interface presents unique opportunities and potential problems related to the long-term viability of the tissue-engineered product. Researchers have been working on understanding and controlling events at these interfaces for years; some of the "classic" studies published ten or twenty years ago are still fresh and very relevant to current research efforts. As more is learned about the clinically desirable and undesirable events that can occur at the tissue-biomaterial interface, new ideas are sparked about designing biomaterial surfaces to control subsequent cell and tissue functions and making novel biomaterials or cell/biomaterial constructs that truly integrate with the body's natural tissues (Chapters 8, 9, and 10 of this text).

Controlling cell-biomaterial interactions is an important goal for the development of tissue-engineered products. However, influencing even the most fundamental cellular functions (such as adhesion or migration) requires an ability to link and utilize concepts from a variety of scientific/engineering and biomedical disciplines. The main purpose of this textbook is, therefore, to provide a

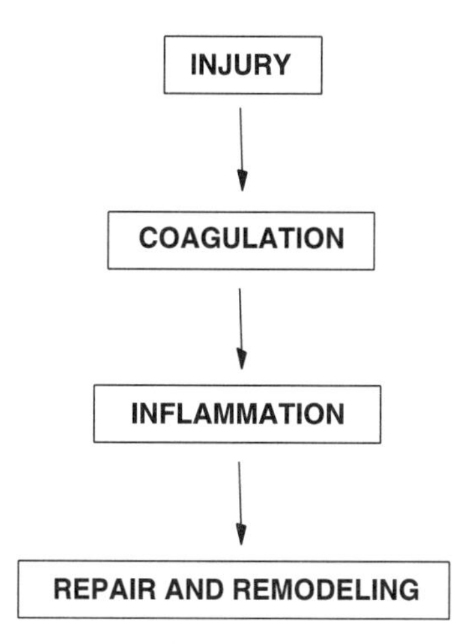

Figure 0.2. Fundamental stages of the wound healing process. This textbook outlines mechanisms by which the wound healing process is initiated by injury and by the presence of synthetic biomaterials, and discusses why these stages are linked in a natural progression. This book also notes some of the major problems that can arise during the wound healing process and how these problems can affect the utility of an implanted material or device.

fundamental, physiology-oriented, and interdisciplinary overview of key concepts in tissue-biomaterial interactions, with the hope of motivating the next generation of scientists and engineers to extend this rapidly growing field of research.

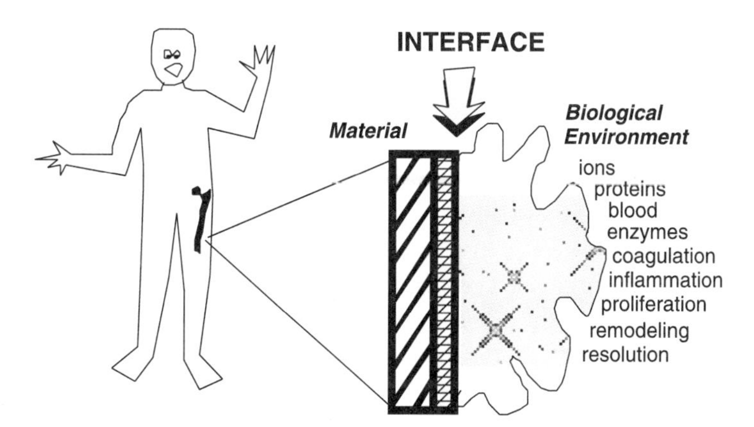

Figure 0.3. The interface between a biomaterial and the body. In this figure, a hip implant is used as an example. Whereas the bulk biomaterial is metallic (titanium, for example) the surface of the implant is probably comprised of a oxide layer (e.g., titanium oxide). The interface between this surface oxide and the biological environment is the location for ions, proteins, enzymes, and other biomolecules to interact with the biomaterial, as well as the location where stages of the body's wound healing processes (Fig. 0.2) will occur.

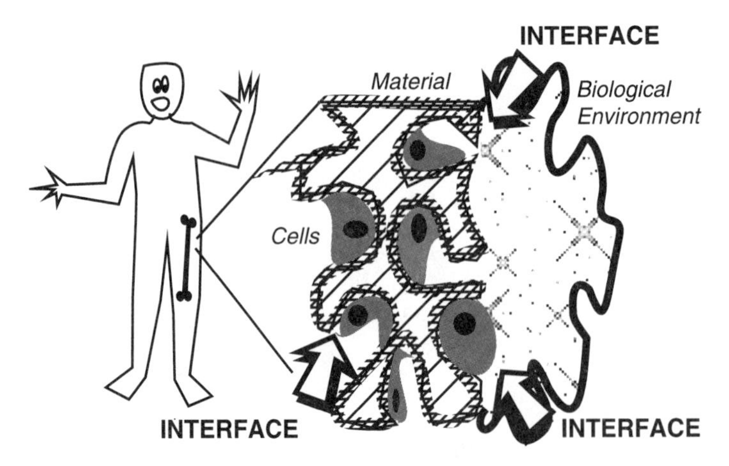

Figure 0.4. The interface(s) between a tissue-engineered product and the body. In this figure, a tissue-engineered femur, consisting of living cells cultured on a biomaterial scaffold, is used as a futuristic example substitution for a traditional metallic hip implant. All of the biological entities and processes noted in Figures 0.1 and 0.2 will still be present at the interface between the material scaffold and the body; but two additional interfaces (between the constituent cells and the material scaffold and between the constituent cells and the body) are present. Designing effective tissue-engineered products requires understanding and controlling events at all three of these interfaces.

SUGGESTED READING

Kasemo, B. and Lausmaa, J., "Surface science aspects of inorganic biomaterials," *CRC Critical Reviews in Biocompatibility*, 2:335–380, 1986.

Williams, D.F., "Tissue-biomaterials interactions," *Journal of Materials Science*, 22:3421–3445, 1987.

1

Biomaterials

1.1 INTRODUCTION

In the past, there was no targeted development of biomaterials based on scientific criteria. Instead, devices consisting of materials that had been designed, synthesized, and fabricated for various industrial needs (for example, the textile, aerospace, and defense industries) were tested in a trial-and-error fashion in the bodies of animals and humans. These unplanned and sporadic attempts had (at best) modest success. Most frequently, the results were unpredictable, mixed, and confounding both in success and in failure.

Because of the continuous and ever-expanding practical needs of medicine and health care practice, there are currently thousands of medical devices, diagnostic products, and disposables on the market. Estimated annual sales of such products in the United States alone are in the order of one hundred billion dollars. In fact, the range of applications continues to grow. In addition to traditional medical devices, diagnostic products, pharmaceutical preparations, and health care disposables, now the list of biomaterial applications includes smart delivery systems for drugs, tissue cultures, engineered tissues, and hybrid organs. To date, tens of millions of people have received medical implants.

Undoubtedly, biomaterials have had a major impact on the practice of contemporary medicine and patient care in both saving, and improving the quality of lives of humans and animals. Modern biomaterial practice still takes advantage of developments in the traditional, nonmedical materials field but is also (actually, more so than ever) aware of, and concerned about, the biocompatibility and biofunctionality of implants.

1.1.1 Definition

Biomaterials is a term used to indicate materials that constitute parts of medical implants, extracorporeal devices, and disposables that have been utilized in medicine, surgery, dentistry, and veterinary medicine as well as in every aspect of patient health care. The National Institutes of Health Consensus Development Conference defined a biomaterial as "any substance (other than a drug) or combination of substances, synthetic or natural in origin, which can be used for any period of time, as a whole or as a part of a system which treats, augments, or replaces any tissue, organ, or function of the body" (Boretos and Eden, 1984). The common denominator in all the definitions that have been proposed for "biomaterials" is the undisputed recognition that biomaterials are distinct from other classes of materials because of the special biocompatibility criteria they must meet. The biocompatibility aspects of biomaterials are addressed in Chapter 9.

Admittedly, any current definition of biomaterials is neither perfect nor complete but has provided an excellent reference or starting point for discussion. It was inevitable that such a definition would need updating to reflect both the evolution of, and revolution in, the dynamic biomedical field. For example, there is an increased emphasis on developing nontraditional clinical methodologies, such as preventing and curing major genetic diseases. These trends in medicine present unique challenges for the biomaterials field. Applications such as controlled delivery of pharmaceuticals (drugs and vaccines), virally and nonvirally mediated delivery agents for gene therapy, and engineered functional tissues require vision, nontraditional thinking, and novel design approaches. Most importantly, to meet the present and future biomaterials challenges successfully, we need materials scientists and engineers who are familiar with and sensitive to cellular, biochemical, molecular, and genetic issues and who work effectively in teams of professionals who include molecular biologists, biochemists, geneticists, physicians, and surgeons.

Synthetic materials currently used for biomedical applications include metals and alloys, polymers, and ceramics. Because the structures of these materials differ, they have different properties and, therefore, different uses in the body. These three classes of materials are reviewed in the remainder of this chapter.

1.2 METALLIC BIOMATERIALS

Metals have been used almost exclusively for load-bearing implants, such as hip and knee prostheses and fracture fixation wires, pins, screws, and plates. Metals have also been used as parts of artificial heart valves, as vascular stents, and as pacemaker leads. Although pure metals are sometimes used, alloys (metals containing two or more elements) frequently provide improvement in material properties, such as strength and corrosion resistance. Three material groups dominate biomedical metals: 316L stainless steel, cobalt-chromium-molybdenum alloy, and pure titanium and titanium alloys (Table 1.1). The main considerations in selecting metals and alloys for biomedical applications are biocompatibility, appropriate mechanical properties, corrosion resistance, and reasonable cost.

TABLE 1.1. Surgical Implant Alloy Compositions (wt %)

Element	316L Stainless Steel (ASTM F138,139)	Co–Cr–Mo (ASTM F799)	Grade 4 Ti (ASTM F67)	Ti-6Al-4V (ASTM F136)
Al	—	—	—	5.5–6.5
C	0.03 max	0.35 max	0.010 max	0.08 max
Co	—	Balance	—	—
Cr	17.0	26.0–30.0	—	—
Fe	Balance	0.75 max	0.30–0.50	0.25 max
H	—	—	0.0125–0.015	0.0125 max
Mo	2.00	5.0–7.0	—	—
Mn	2.00 max	1.0 max	—	—
N	—	0.25 max	0.03–0.05	0.05 max
Ni	10.00	1.0 max	—	—
O	—	—	0.18–0.40	0.13 max
P	0.03 max	—	—	—
S	0.03 max	—	—	—
Si	0.75 max	1.0 max	—	—
Ti	—	—	Balance	Balance
V	—	—	—	3.5–4.5
W	—	—	—	—

1.2.1 Basis of Structure-Property Relationships

The properties of materials are governed by their structure. At the atomic level, metals consist of positively charged ion cores immersed in a "cloud" of loosely bound electrons. This atomic level structure is responsible for the characteristic and distinct properties of metals. Metallic bonding allows the atoms to organize themselves into an ordered, repeating, three-dimensional crystalline pattern, which can be visualized as the packing of hard spheres into cubic or hexagonal arrangements. The delocalized electrons are responsible for the electrical and thermal conductivity of metals. Because the interatomic bonds are not spatially directed in metals, planes of atoms can "slip" over one another to allow plastic (permanent) deformation.

The chemical properties of materials also are related to the nature of their atomic bonding. The more resistant the constituent atoms/ions are to being separated, the more inert the material will be. In metals, the loose, nondirectional way in which the electrons are bonded allows the atoms/ions to be parted more easily. Consequently, although their mechanical properties make metals the appropriate choice for many biomedical applications, susceptibility to chemical degradation is an aspect that must be considered.

Because the interactions between cells and tissues with biomaterials at the tissue implant interface are almost exclusively surface phenomena, surface properties of implant materials are of great importance. A surface is the termination of the normal three-dimensional structure of a material. Lack of near neighbor atoms on one side of the surface alters the electronic structure and consequently the way these atoms interact with other atoms. Chemical bonds will "dangle" into the space outside the solid material and will result in the surface atoms having higher energy than do atoms in the bulk. As a result, surface atoms will attempt to reduce free energy by rearranging and/or bonding to any available reactive molecules to reach a more favorable energy state.

1.2.2 Corrosion

The physiological environment is typically modeled as a 37 °C aqueous solution, at pH 7.3, with dissolved gases (such as oxygen), electrolytes, cells, and proteins. Immersion of metals in this environment can lead to corrosion, which is deterioration and removal of the metal by chemical reactions. During the electrochemical process of corrosion, metallic biomaterials can release ions, which may reduce the biocompatibility of materials and jeopardize the fate of implants. For example, the type and concentration of released corrosion products can alter the functions of cells in the vicinity of implants as well as of cells at remote locations after transport of the corrosion by-products to distant sites inside the body. These circumstances become stronger possibilities in the bodies of sick and elderly patients, who are the largest group of recipients of prostheses.

Even before implantation, through chemical reaction of metals with the oxygen in ambient air or by oxidation in an acidic solution, an oxide surface film forms on their surface. Because oxides are ceramics (see Section 1.3), which are electrical and thermal insulators, the electrochemical reactions that lead to corrosion are reduced or prevented. In other words, the oxidized metallic surfaces are "passivated." In fact, the stability of the oxides present on different metals determines their overall corrosion resistance. For example, even though 316L stainless steel implants perform satisfactorily in short-term applications, such as fracture fixation, they are susceptible to crevice corrosion and pitting when implanted for longer periods. Titanium and its alloys, as well as cobalt-chromium alloys, have more favorable corrosion resistance for long-term implant applications such as joint and dental prostheses.

1.2.3 Mechanical Properties

The mechanical properties of materials are of great importance when designing load-bearing orthopedic and dental implants. Some mechanical properties of metallic biomaterials are listed in Table 1.2. With a few exceptions, the high tensile and fatigue strength of metals, compared with ceramics and polymers, make them the materials of choice for implants that carry mechanical loads.

It should be noted that, in contrast to the nanophase, composite nature of

TABLE 1.2. Select Properties of Metallic Biomaterials*

Material	Young's Modulus, E (GPa)	Yield Strength, σ_y (MPa)	Tensile Strength, σ_{UTS} (MPa)	Fatigue Limit, σ_{end} (MPa)
Stainless steel	190	221–1,213	586–1,351	241–820
Cobalt-chromium (Co–Cr) alloys	210–253	448–1,606	655–1,896	207–950
Titanium (Ti)	110	485	760	300
Ti-6Al-4V	116	896–1,034	965–1,103	620
Cortical bone	15–30	30–70	70–150	

*Adapted from J.B. Brunski. Metals, pp. 37–50 in B.D. Ratner, A.S. Hoffman, F.J. Shoen, and J.E. Lemons (eds.), *Biomaterials Science: An Introduction to Materials in Medicine*, Academic Press, San Diego (1996).

tissue such as bone, the biomedical metals used for implants are conventional, homogeneous materials. The elastic moduli of the metals listed in Table 1.2 are at least seven times greater than that of natural bone. This mismatch of mechanical properties can cause "stress shielding," a condition characterized by bone resorption (loss of bone) in the vicinity of implants. This clinical complication arises because preferential distribution of mechanical loading through the metallic prosthesis deprives bone of the mechanical stimulation needed to maintain homeostasis.

The mechanical properties of a specific implant depend not only on the type of metal but also on the processes used to fabricate the material and device. Thermal and mechanical processing conditions can change the microstructure of materials. For example, in "cold-working" a metal, such as by rolling or forging, the resulting deformation makes the material stronger and harder. Unfortunately, as the metal becomes harder and stronger it also becomes less ductile (undergoes less deformation before failure) and more chemically reactive.

Compared with the elastic moduli of either stainless steel or cobalt-chromium molybaenum alloys, Ti and Ti-6Al-4V have much lower (approximately half) moduli that are still almost an order of magnitude higher than that of bone. Another advantage of Ti-based metals as a bone implant material is their favorable strength-to-density ratio. Stainless steel and Co–Cr alloys have densities of approximately 8.8 g/cm^3 and 7.8 g/cm^3, respectively. Because Ti has a density of only 4.5 g/cm^3, its strength-to-density ratio is larger. Disadvantages of titanium for medical use include a relatively low shear strength, poor wear resistance, and difficulties in fabrication. The stable, coherent titanium oxide (TiO_2) film that forms on titanium and its alloys gives them superior corrosion resistance compared with stainless steel and Co–Cr alloys. The oxidized surface is also believed to be responsible for Ti implants becoming osseointegrated in vivo, a process whereby bone is aposed to the implant without chronic inflammation and without an intervening fibrous capsule.

TABLE 1.3. Ceramics Used in Biomedical Applications

Ceramic	Chemical Formula	Comment
Alumina	Al_2O_3	Bioinert
Zirconia	ZrO_2	
Pyrolytic carbon		
Bioglass	$Na_2OCaOP_2O_3–SiO$	Bioactive
Hydroxyapatite (sintered at high temperature)	$Ca_{10}(PO_4)_6(OH)_2$	
Hydroxyapatite (sintered at low temperature)	$Ca_{10}(PO_4)_6(OH)_2$	Biodegradable
Tricalcium phosphate	$Ca_3(PO_4)_2$	

Definitions:

Bioinert refers to a material that retains its structure in the body after implantation and does not induce any immunologic host reactions.

Bioactive refers to materials that form bonds with living tissue.

Biodegradable refers to materials that degrade (by hydrolytic breakdown) in the body while they are being replaced by regenerating natural tissue; the chemical by-products of the degrading materials are absorbed and released via metabolic processes of the body.

1.3 CERAMIC AND GLASS BIOMATERIALS

Ceramics and glasses are used as components of hip implants, dental implants, middle ear implants, and heart valves. Overall, however, these biomaterials have been used less extensively than either metals or polymers. Some ceramics that have been used for biomedical applications are listed in Table 1.3.

1.3.1 Basis of Structure-Property Relationships

Ceramics are materials composed of metallic and nonmetallic elements held together by ionic and/or covalent bonds. As with metals, the interatomic bonds in ceramics result in long-range three-dimensional crystalline structures; glasses do not have long-range order. In contrast to metallic bonding, the electrons in ionic and covalent bonds are localized between the constituent ions/atoms. Consequently, ceramics are typically electrical and thermal insulators. The strong ionic and covalent bonds also make ceramics hard and brittle, because the planes of atoms/ions cannot slip past one another. Ceramics and glasses typically fail with little, if any, plastic deformation, and they are sensitive to the presence of cracks or other defects. The ionic and/or covalent nature of ceramics also influences their chemical behavior.

1.3.2 Degradation

Although they do not undergo corrosion, ceramics and glasses are susceptible to other forms of degradation when exposed to the physiological environment. The

TABLE 1.4. Mechanical Properties of Ceramic Biomaterials*

	Young's Modulus, E (GPa)	Compressive Strength, σ_{UCS} (MPa)	Tensile Strength, σ_{UTS} (MPa)
Alumina	380	4500	350
Bioglass-ceramics	22	500	56–83
Calcium phosphates	40–117	510–896	69–193
Pyrolytic carbon	18–28	517	280–560

*Compiled from L.L. Hench. Ceramics, Glasses, and Glass-Ceramics, pp. 73–84 in B.D. Ratner, A.S. Hoffman, F.J. Shoen, and J.E. Lemons (eds), *Biomaterials Science: An Introduction to Materials in Medicine*, Academic Press, San Diego (1996); J.B. Park and R.S. Lakes, *Biomaterials*, Plenum Press, New York (1992); and J. Black, *Biological Performance of Materials*, Marcel Dekker, New York (1992).

mechanism and rate of degradation, however, depend on the particular type of ceramic. Even alumina, which is generally considered a bioinert ceramic, experiences a time-dependent decrease in strength during immersion in saline in vitro and after implantation. This process may result from a preferential dissolution of impurities that results in crack propagation. Bioactive ceramics and glasses are also degraded in the body. Not only can they undergo slow or rapid dissolution (depending on the composition and processing history of the material), but because of the similarity of calcium phosphates to the mineral component of bone, they may also be resorbed by osteoclasts (the cells that break down bone).

1.3.3 Mechanical Properties

The major drawbacks to the use of ceramics and glasses as implants are their brittleness and poor tensile properties (Table 1.4). Although they can have outstanding strength when loaded in compression, ceramics and glasses fail at low stress when loaded in tension or bending. Among biomedical ceramics, alumina has the highest mechanical properties, but its tensile properties are still below those of metallic biomaterials. Additional advantageous properties of alumina are its low coefficient of friction and wear rate. Because of these properties, alumina has been used as a bearing surface in joint replacements.

The mechanical properties of calcium phosphates and bioactive glasses make them unsuitable as load-bearing implants. Clinically, hydroxyapatite has been used as a filler for bone defects and as an implant in load-free anatomic sites (for example, nasal septal bone and middle ear). In addition, hydroxyapatite has been used as a coating on metallic orthopedic and dental implants to promote their fixation in bone. In these cases, the underlying metal carries the load, whereas the surrounding bone strongly bonds to hydroxyapatite. Delamination of the ceramic layer from the metal surface, however, can create serious problems and lead to implant failure.

TABLE 1.5. Examples of Biomedical Applications of Polymers

Applications	Polymer(s)
Cardiovascular implants	Polyethylene; poly(vinyl chloride); polyester; silicone rubber; poly(ethylene terephthalate); polytetrafluoroethylene
Orthopedic implants	Ultra-high-molecular-weight polyethylene; poly(methyl methacrylate)
Drug release	Poly(lactide-co-glycolide)
Tissue engineering	Poly(lactic acid); poly(glycolic acid); poly(lactide-co-glycolide)

1.4 POLYMERIC BIOMATERIALS

Polymers are the most widely used materials in biomedical applications. They are the materials of choice for cardiovascular devices as well as for replacement and augmentation of various soft tissues. Polymers also are used in drug delivery systems, in diagnostic aids, and as a scaffolding material for tissue engineering applications. Examples of current applications include vascular grafts, heart valves, artificial hearts, breast implants, contact lenses, intraocular lenses, components of extracorporeal oxygenators, dialyzers and plasmapheresis units, coatings for pharmaceutical tablets and capsules, sutures, adhesives, and blood substitutes. Examples of polymers and their uses are given in Table 1.5.

1.4.1 Basis of Structure-Property Relationships

Polymers are organic materials consisting of large macromolecules composed of many repeating units (called "mers"). These long molecules are covalently bonded chains of atoms. Unless they are cross-linked, the macromolecules interact with one another by weak secondary bonds (hydrogen and van der Waals bonds) and by entanglement. Because of the covalent nature of interatomic bonding within the molecules, the electrons are localized, and consequently polymers tend to be poor thermal and electric conductors.

The mechanical and thermal behavior of polymers is influenced by several factors, including the composition of the backbone and side groups, the structure of the chains, and the molecular weight of the molecules. Plastic deformation occurs when the applied mechanical forces cause the macromolecular chains to slide past one another. Changes in polymer composition or structure that increase resistance to relative movement of the chains increase the strength and decrease the plasticity of the material. Substitutions into the backbone that increase its rigidity hinder movement of the chains. Bulky side groups also make disentanglement more difficult. Increasing macromolecule length (molecular weight) also makes the chains less mobile and hinders their relative movement.

1.4.2 Degradation

Degradation of polymers requires disruption of their macromolecular structure and can occur by either alteration of the covalent interatomic bonds in the chains or alteration of the intermolecular interactions between chains. The former can occur by chain scission (cleavage of chains) or cross-linking (joining together of adjacent chains), an unlikely occurrence under physiological conditions. The latter can occur by incorporation (absorption) or loss (leaching) of low-molecular-weight compounds. As described in Chapter 8, Section 8.2, chemical reactions, such as oxidation and hydrolysis, can also change the properties of implanted polymers. For polymers, the method of sterilizing the biomaterial can significantly alter its properties. For example, high temperatures (121–180 °C), steam, chemicals (ethylene oxide), and radiation can compromise the shape and/or mechanical properties of polymeric materials.

Polymers may contain various (often unspecified) additives, traces of catalysts, inhibitors, and other chemical compounds needed for their synthesis. Over time in the physiological environment, these compounds can leach from the polymer surface. As is the case with corrosion by-products released from metallic implants, the chemicals released from polymers may induce adverse local and systemic host reactions that cause clinical complications. This release is a concern for materials, such as bone cement, that are polymerized in the body and for flexible polymers, such as poly(vinyl chloride), that contain low-molecular-weight species (plasticizers) to make them pliable.

In addition to unintentional degradation, certain polymers have been designed to undergo controlled degradation. Among biodegradable polymers, poly(lactic acid), poly(glycolic acid), and their copolymers have been the most widely used. These materials degrade into smaller fragments as well as monomers, such as lactic acid, that can be eliminated by normal metabolic processes of the body. Biodegradable polymers are used for sutures, controlled drug delivery, tissue engineering, and fracture fixation.

1.4.3 Mechanical Properties

The mechanical properties of polymers depend on several factors, including the composition and structure of the macromolecular chains and their molecular weight. Table 1.6 lists some mechanical properties of selected polymeric biomaterials. Compared with metals and ceramics, polymers have much lower strengths and moduli but they can be deformed to a greater extent before failure. Consequently, polymers are generally not used in biomedical applications that bear loads (such as body weight). Ultra-high-molecular-weight polyethylene is an exception, as it is used as a bearing surface in hip and knee replacements. The mechanical properties of polymers, however, are sufficient for numerous biomedical applications (some of which are listed in Table 1.5).

TABLE 1.6. Mechanical Properties of Polymers*

Polymer	Tensile Strength σ_{UTS} (MPa)	Young's Modulus, E (GPa)	% Elongation
Poly(methyl methacrylate) (PMMA)	30	2.2	1.4
Nylon 6/6	76	2.8	90
Poly(ethylene terephthalate)	53	2.14	300
Poly(lactic acid)	28–50	1.2–3	2–6
Polypropylene	28–36	1.1–1.55	400–900
Polytetrafluoroethylene	17–28	0.5	120–350
Silicone rubber	2.8	Up to 10	160
Ultra-high-molecular-weight polyethylene (UHMWPE)	≥35	4–12	≥300

*Compiled from J. Kohn and R. Langer. Bioresorbable and Bioerodible Materials, pp. 64–73 in B.D. Ratner, A.S. Hoffman, F.J. Shoen, and J.E. Lemons (eds.), *Biomaterials Science: An Introduction to Materials in Medicine*, Academic Press, San Diego (1996); and J.B. Park and R.S. Lakes, *Biomaterials*, Plenum Press, New York (1992).

1.5 CHOICE OF MATERIALS FOR BIOMEDICAL APPLICATIONS

In the past, success of materials in biomedical applications was not so much the outcome of meticulous selection based on biocompatibility criteria but rather the result of serendipity, continuous refinement in fabrication technology, and advances in material surface treatment.

In the present and future, election of a biomaterial for a specific application must be based on several criteria. The physicochemical properties and durability of the material, the desired function of the prosthesis, the nature of the physiological environment at the organ/tissue level, adverse effects in case of failure, as well as cost and production issues must be considered for each specific application. Biocompatibility (addressed in Chapter 9) is the paramount criterion that must be met by every biomaterial.

Mechanical requirements must also be taken into consideration when choosing materials for biomedical applications. Material strength (tensile or compressive), stiffness, fatigue endurance, wear resistance, and dimensional stability should be considered with respect to the end use of the prosthetic device to ensure its success. For example, a rigid, strong material would be more suitable for a hip implant, whereas a flexible, less strong material would be sufficient for a vascular graft. Moreover, the performance of materials under dynamic loading conditions must be considered when appropriate, because many implants are subjected to various types and magnitudes of repeated stresses in the body. Consider a hip, knee, or ligament replacement that will be subjected to approximately one million steps per year, while various other physical activities will exert different loads across the joints. At 70 beats per minute, a prosthetic heart valve would experience over three and a half million cycles per year. Other physical properties (such as electrical and thermal conductivity, light transmission, and radiopacity) are important for specific applications, such as pacemaker electrodes, intraocular lenses, and dental restoratives, and must be considered when applicable.

Because the practice of medicine and surgery requires sterile products, decisions regarding choice of biomaterial(s) for a specific application should include consideration of sterilization of the final product(s). Moist heat and high pressure (typical conditions in steam autoclaves), ethylene oxide gas, and gamma radiation are procedures commonly used in sterilizing biomedical materials and devices. Special care should be taken with polymers that do not tolerate heat, absorb and subsequently release ethylene oxide (a toxic substance), and degrade when exposed to radiation.

1.6 BIOMATERIALS FOR IMPLANTABLE DEVICES: PRESENT AND FUTURE DIRECTIONS

Unquestionably, important advances have been made in the clinical use of medical implants and other devices. Presently, emphasis is placed on the design of proactive biomaterials, that is, materials that elicit specific, desired, and timely responses from surrounding cells and tissues. Medical research continues to explore new scientific frontiers for diagnosing, treating, curing, and preventing diseases at the molecular/genetic level. With this newfound knowledge, there will be further need for innovative formulations and/or modifications of existing materials (see Chapter 8, Section 8.4), for novel materials, and for nontraditional applications of biomaterials, such as in tissue engineering. Promising developments include bioinspired chemical and topographic modifications of materials surfaces, current-conducting polymers, and nanophase materials. In addition to new challenges and opportunities, some of the unresolved issues (primarily, biocompatibility) of the past and present will also need to be addressed in the future.

1.7 SUMMARY

- A biomaterial is any substance (other than drugs), natural or synthetic, that treats, augments, or replaces any tissue, organ, and body function.
- The properties of materials are governed by their structure, determined by the way their constituent atoms are bonded together.
- Lack of near neighbor atoms, caused by creation of a surface, results in different surface versus bulk material properties that have major consequences for tissue-implant interactions.
- The mechanical properties (e.g., strength, modulus, and fatigue limit) of metals makes them desirable choices for many load-bearing biomedical prostheses applications.

- Metals are susceptible to degradation by corrosion, a process that can release by-products (such as ions, chemical compounds, and particulate debris) that may cause adverse biological responses.
- Ceramics are attractive biomaterials because they can be either bioinert, bioactive, or biodegradable; however, they have serious drawbacks because they are brittle and have low tensile strength.
- The properties of polymers depend on the composition, structure, and arrangement of their constituent macromolecules.

1.8 BIBLIOGRAPHY/SUGGESTED READING

Alexander H., Brunski J.B., Cooper S.L., Hench L.L., Hergenrother R.W., Hoffman A.S., Kohn J., Langer R., Peppas N.A., Ratner B.D., Shalaby S.W., Visser S.A., and Yannas I.V., Classes of materials used in medicine. In *Biomaterials Science: An Introduction to Materials in Medicine*, Ratner, B.D., Hoffman, A.S., Schoen, F.J., Lemons, J.E., (eds.) Academic Press, New York, NY (1996), pp. 37–130.

Boretos, J.W., Eden, M. *Contemporary Biomaterials, Material and Host Response, Clinical Applications, New Technology and Legal Aspects.* Noyes Publications, Park Ridge, NJ (1984), pp. 232–233.

Cooke F.W., Lemons J.E., and Ratner B.D., Properties of Materials in *Biomaterials Science: An Introduction to Materials in Medicine*, Ratner, B.D., Hoffman, A.S., Schoen, F.J., Lemons, J.E., (eds.) Academic Press, New York, NY (1996), pp. 11–35.

Peppas, N.A., Langer, R. "New Challenges in Biomaterials", *Science*: 263 (1994), pp. 1715–1720.

Ratner, B.D. "New ideas in biomaterials science—a path to engineered biomaterials." *Journal of Biomedical Materials Research*: 27 (1993), pp. 837–850.

1.9 QUIZ QUESTIONS

1. Define the term "biomaterial."

2. Define the term "metal." Give an example of a metal and describe its use in a biomedical prosthesis.

3. Define the term "ceramic." Give an example of a ceramic and describe its use in a biomedical prosthesis.

4. Define the term "polymer." Give an example of a polymer and describe its use in a biomedical prosthesis.

5. Why does the method of processing a material change its properties?

6. What are the potential consequences of a biomaterial degrading after implantation of a prosthetic device?

1.10 STUDY QUESTIONS

1. You are the biomedical engineer in charge of a project to design a novel material for bone replacement.
   (a) What synthetic material(s) will you choose for this application? Justify your choice(s).
   (b) What criteria should the new material satisfy?
   (c) What are the advantages of the new material?
   (d) What are the disadvantages of the new material?

2. Repeat Study Question #1, but now consider the design of a novel material for replacement of diseased blood-vessel wall tissue (vascular grafts).

3. What are the advantages and disadvantages of using bioactive and biodegradable ceramics in bone replacement?

4. Discuss how biomaterials in the next century may differ from those presently used for biomedical applications.

2

Proteins

2.1 INTRODUCTION

Proteins perform a large number of functions in the body. For example, they serve a structural role in providing the supporting matrix of many tissues. As enzymes, they catalyze thousands of important chemical reactions essential to life, from regulation of gene transcription to conversion of food to energy. Changes in the levels of proteins or mutations in their structure, which lead to altered function, are responsible for many diseases. As discussed in other chapters, proteins also play an important role in determining the nature of the tissue-implant interface.

2.2 PRIMARY STRUCTURE

Amino acids are the building blocks of proteins. As shown in Figure 2.1, the general structure of the common amino acids consists of a central α-carbon atom to which a carboxyl group, an amino group, a hydrogen atom, and a specific functional side chain (R) are attached. Depending on pH, the carboxyl and amino groups can be charged (Fig. 2.2). At neutral pH, amino acids are zwitterions, meaning that they possess equal amounts of positive and negative charge, because the carboxylic acid group is unprotonated and the amino group is protonated.

There are 20 amino acids for which at least one specific codon (genetic coding sequence) exists. Depending on the side chain, amino acids can be divided into four groups: nonpolar (hydrophobic), polar, positively charged, or negatively

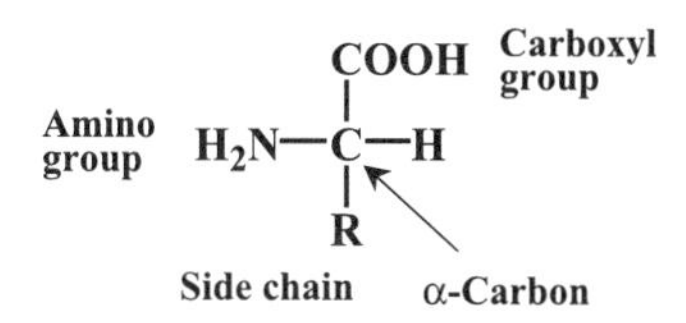

Figure 2.1. General structure of an amino acid with the important functional groups labeled.

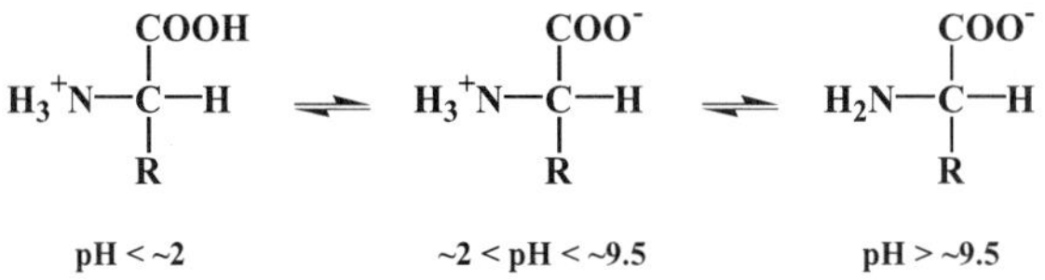

$$H_3^+N{-}C(COOH)(R){-}H \rightleftharpoons H_3^+N{-}C(COO^-)(R){-}H \rightleftharpoons H_2N{-}C(COO^-)(R){-}H$$

pH < ~2 ~2 < pH < ~9.5 pH > ~9.5

Figure 2.2. pH dependence of charge of an amino acid.

charged (Fig. 2.3). As described for the carboxyl and amino groups on the α-carbon, the side chains of charged amino acids can contain functional groups that exhibit pH-dependent ionization. For example, amino acids whose side chains possess amide moieties can be protonated to give a positive charge, and carboxyl-terminated side chains can be deprotonated to yield a negative charge. Proline is unique in that it is cyclic, with the side chain covalently bonded to the nitrogen of the amino group. Technically, it is an imino acid rather than an amino acid. The cyclic structure limits rotational flexibility of the nitrogen-carbon bond and consequently affects protein structure. To simplify representation of amino acid sequences, three-letter and one-letter abbreviations have been established (Table 2.1).

In addition to the 20 common amino acids, proteins can contain derived amino acids, which are the result of enzymatic actions on amino acids already incorporated into the protein. Examples include hydroxylated derivatives of proline and lysine, a phosphorylated derivative of serine, and a carboxylated derivative of glutamic acid (Fig. 2.4).

Proteins are formed by polymerization of amino acids in enzyme-catalyzed condensation reactions (Fig. 2.5). The amino group of one peptide (amino acid) reacts with the carboxyl group of a second peptide, eliminating a water molecule and resulting in a peptide (amide) bond. Polypeptides are formed by sequential addition of amino acids. A molecule consisting of two amino acids is called a dipeptide, three amino acids a tripeptide, four amino acids a tetrapeptide, etc. The exact linear sequence of amino acids is the primary structure (Fig. 2.6).

A vast number of primary structures can be constructed from the 20 common amino acids. For even a short polypeptide containing 25 amino acids, 20^{25} combinations are possible. Because the different amino acids with their different side chains permit different types of chemical bonding, each polypeptide will have different higher levels of protein structure and therefore function. Although some substitutions (such as replacing a polar peptide with a polar peptide of similar size) in the primary structure have minimal effect, changes in sequence can significantly alter protein structure. For example, sickle cell anemia is caused by substitution of valine for glutamic acid in hemoglobin, the protein that carries oxygen in blood. The altered hemoglobin molecules (called hemoglobin S) resulting from the mutation create malformed red blood cells, which clog blood vessels and restrict the supply of blood to tissues.

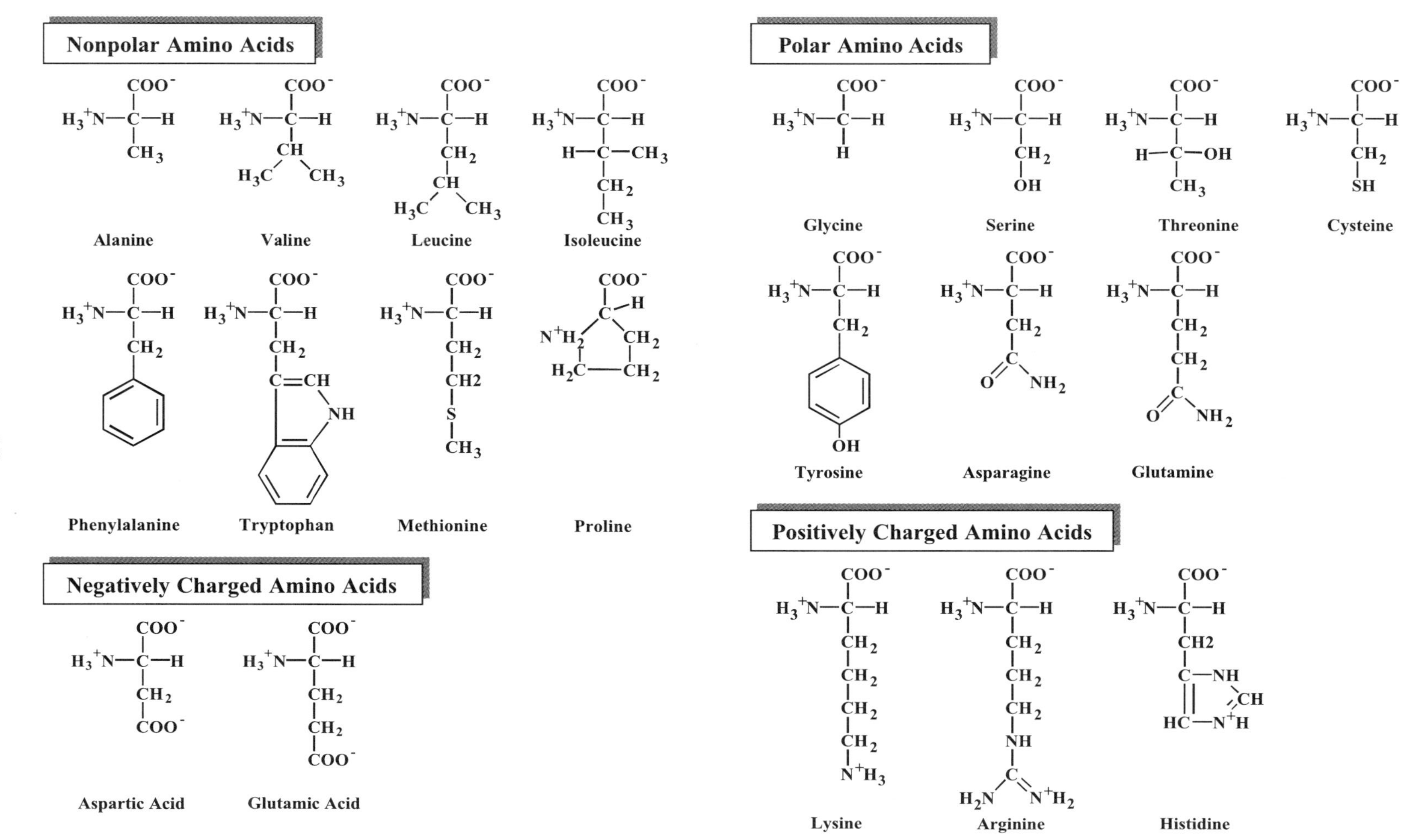

Figure 2.3. The 20 common amino acids.

TABLE 2.1. One- and Three-Letter Abbreviations for the 20 Common Amino Acids

Amino Acid	Three Letter	One Letter
Alanine	Ala	A
Cysteine	Cys	C
Aspartic Acid	Asp	D
Glutamic Acid	Glu	E
Phenylalanine	Phe	F
Glycine	Gly	G
Histidine	His	H
Isoleucine	Ile	I
Lysine	Lys	K
Leucine	Leu	L
Methionine	Met	M
Asparagine	Asn	N
Proline	Pro	P
Glutamine	Gln	Q
Arginine	Arg	R
Serine	Ser	S
Threonine	Thr	T
Valine	Val	V
Tryptophan	Trp	W
Tyrosine	Tyr	Y

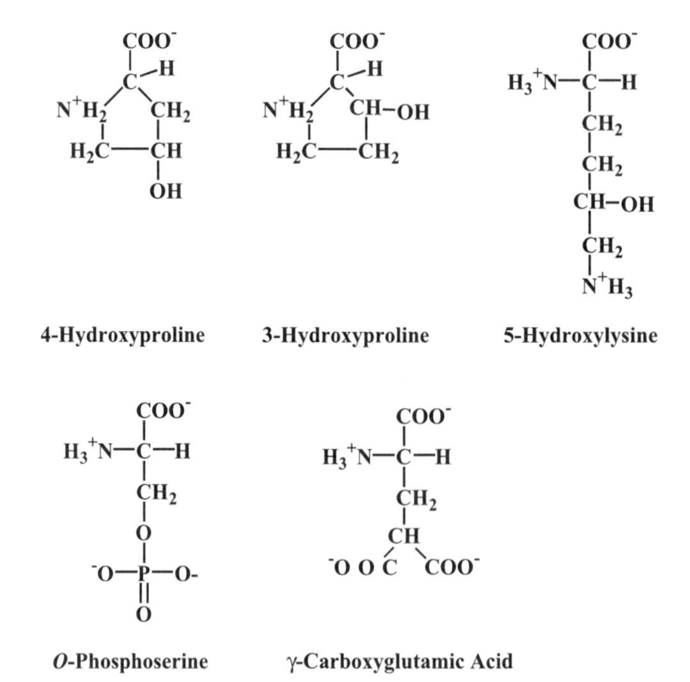

Figure 2.4. Examples of derived amino acids obtained by modification of a common amino acid after incorporation into a protein.

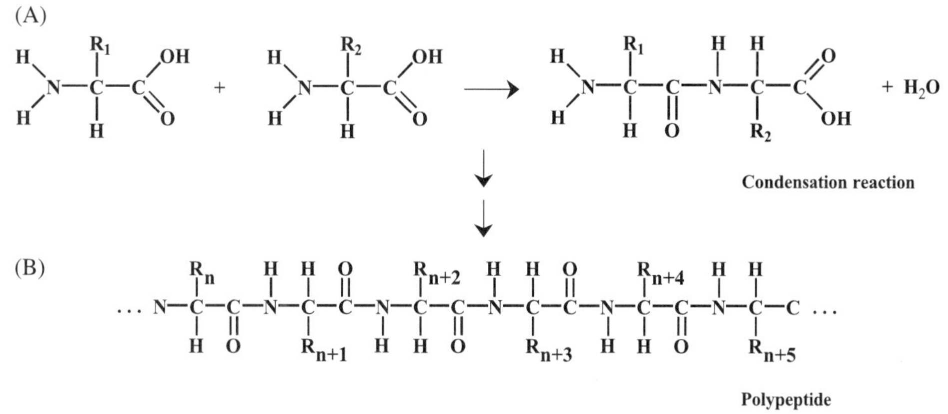

(A)

Condensation reaction

(B)

Polypeptide

Figure 2.5. Formation of a polypeptide. **(A)** Condensation reaction resulting in a peptide bond. **(B)** Sequential addition of amino acids resulting in a polypeptide.

2.3 SECONDARY STRUCTURE

Proteins do not exist simply as long, extended chains of amino acids. Instead, interactions between amino acids cause folding, bending, and coiling of the chain to give a specific three-dimensional structure (conformation). Consider a telephone cord. Rather than being straight, the cord is a right-handed helix, and when randomly dropped on a table, it has a more complex arrangement, with the helix crisscrossing over itself.

Secondary structure refers to interactions within localized domains that result in three-dimensional arrangement. In the example of a telephone cord, secondary structure is represented by the helical arrangement of what would otherwise be a straight piece of plastic. At this level, coiling and bending are caused by hydrogen bonding between a carbonyl group (–C=O) of one peptide bond and a secondary amine (–NH) of another. Bending of the polypeptide is constrained by the rotational angles of the covalent bonds in the chain. Because the peptide bond has partial double bond character, atoms attached to the carbonyl carbon and the nitrogen lie in a common plane. Consequently, the polypeptide chain acts like a chain of plates connected at the α-carbon atoms (Fig. 2.7).

Although many secondary structures are possible, the α-helix and β-structures are the most thermodynamically stable. In the α-helix, each peptide forms hydrogen bonds with the fourth amino acid above it and the fourth amino acid below it (Fig. 2.8). The hydrogen bonds are parallel to the axis of the helix. Helices with either left- or right-handed spiral character can be formed, but the right-handed helix is more stable. Technically, there are 3.6 amino acids per turn of the α-helix. Dipoles in the helix are aligned, meaning that the C=O groups point in one direction and the NH groups point in the opposite direction. As shown in Figure 2.8, the side chains of the amino acids are on the out-

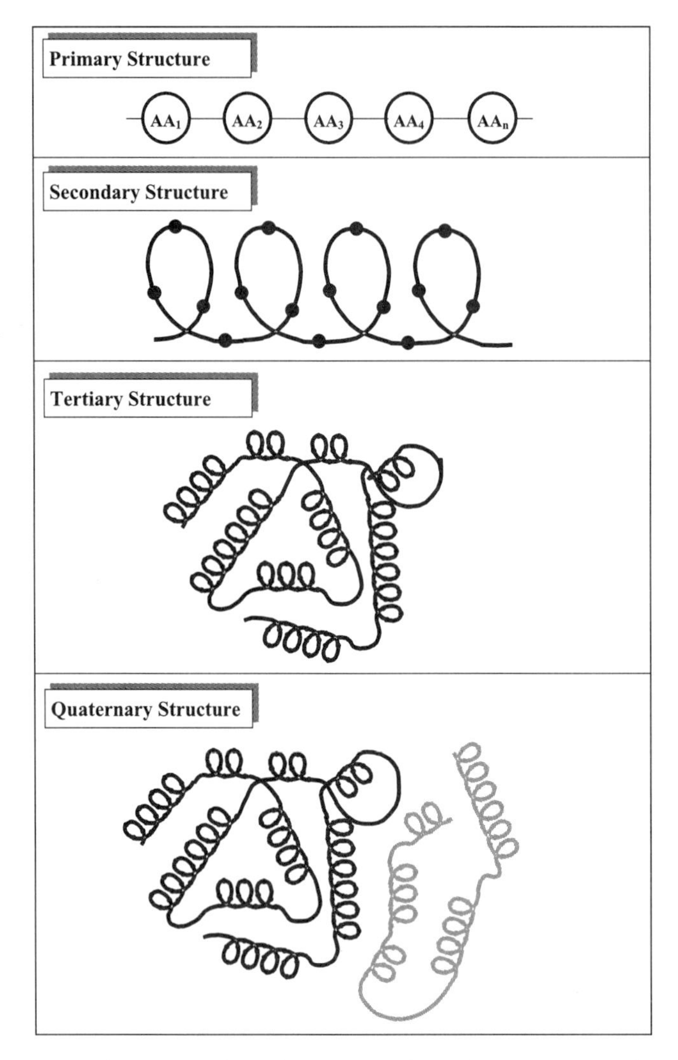

Figure 2.6. Schematic illustration of the four levels of protein structure. [Adapted from Kleinsmith, L.J. and Kish, V.M., *Principles of Cell Biology*, Harper & Row, New York, 1988.]

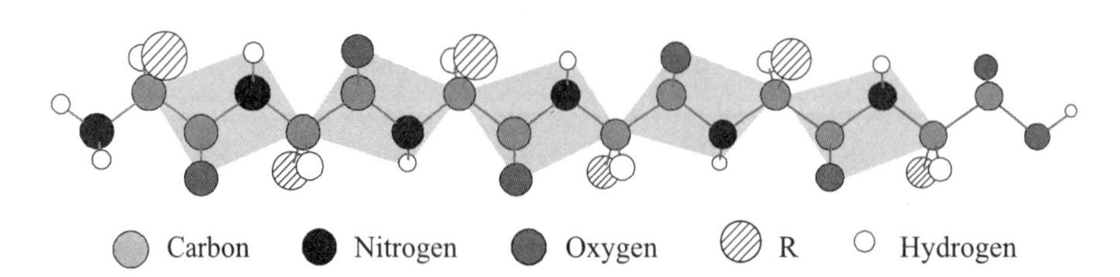

Figure 2.7. The C–N bond constrains rotation, causing the polypeptide chain to behave as a chain of plates.

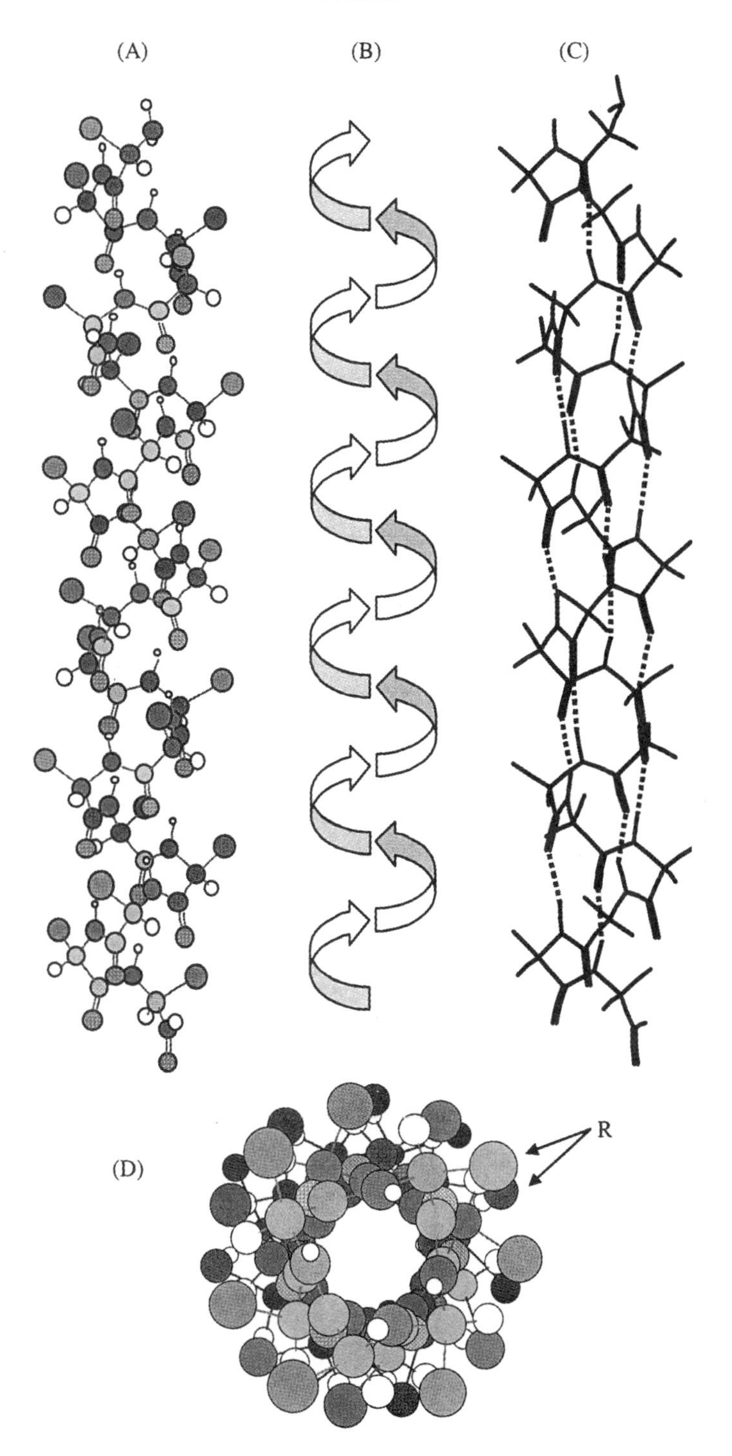

Figure 2.8. The α-helix. (**A**) Ball and stick model. (**B**) Arrows showing the direction of rotation, with shaded sections representing regions behind the plane of the page. (**C**) Simplified stick model showing H bonds (dashed lines) that stabilize the helix. (**D**) Axial view showing side chains on the outside of the helix.

side of the helix, which enables them to easily interact with those in other domains of the protein or in other biomolecules.

In addition to coils, polypeptides can exist as a extended zigzag chains. A region of polypeptide with an extended chain conformation is called a β-strand. β-Sheets are stabilized by hydrogen bonds between two or more β-strands (Fig. 2.9). A parallel β-sheet forms when each section of the chain runs in the same direction. An antiparallel structure is formed when the polypeptide chain folds back and forth on itself, with each section of the chain running in the opposite direction. These structures have the appearance of a pleated sheet (Fig. 2.10). The side chains project above and below the planes of the β-sheet.

Regions of other regular or random structure will be intermixed with α-helical and β-structures throughout a protein (Fig. 2.11). For example, helices deviating from the α-helix can result from the presence of the cyclic amino acid proline, which sterically prevents the N atoms from forming hydrogen bonds that stabilize the α-helical structure. Charge repulsion resulting from too many amino acids of like charge also can cause a more random configuration.

2.4 TERTIARY STRUCTURE

Tertiary structure refers to the three-dimensional structure of the whole protein subunit (Fig. 2.6). Again using the example of a telephone cord, tertiary structure is represented by the folding of the coiled plastic back and forth. Whereas secondary structure is governed by localized interactions between amino acids, tertiary structure is based on interactions between distant sections of the polypeptide chain. Hydrogen bonding between carboxyl and secondary amino groups in the backbone is the basis of secondary structure, but interaction between side chains of amino acids is important for tertiary structure. Depending on the particular amino acids involved, at least four types of chemical interactions can occur between side chains. Covalent disulfide bonds can form between cysteine residues. Ionic interaction can occur between positively charged side chains (lysine, arginine, and histidine) and negatively charged side chains (aspartic acid and glutamic acid). Hydrogen bonds can occur between polar amino acids, such as serine and tyrosine, and other amino acids. Hydrophobic interactions involve nonpolar amino acids, such as leucine, phenylalanine, tryptophan, and valine. Rather than a direct attractive or repulsive effect between amino acids, hydrophobic interactions are based primarily on the mutual "water-hating" (hydrophobic) nature of the side chains; nonpolar groups are forced together because of their inability to interact with water. Combined, these various interactions cause twisting and folding of the polypeptide chain.

Protein folding attempts to maximize exposure of polar groups to the aqueous environment while minimizing exposure of hydrophobic groups. Thus polypeptide chains are generally folded to place hydrophobic side chains toward the

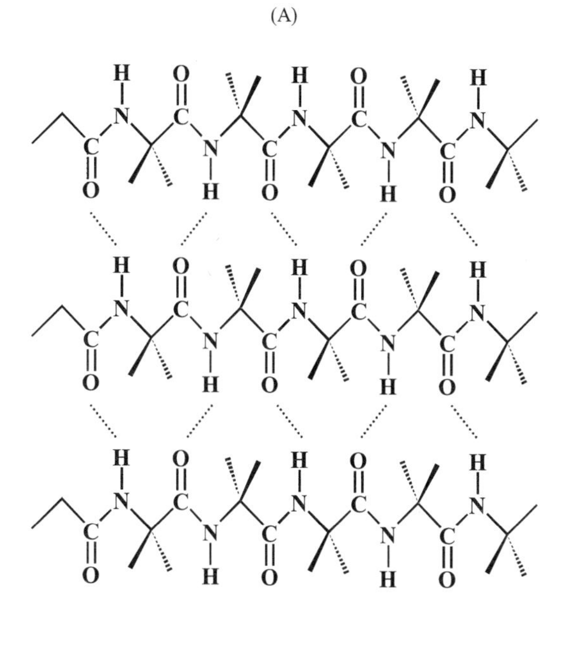

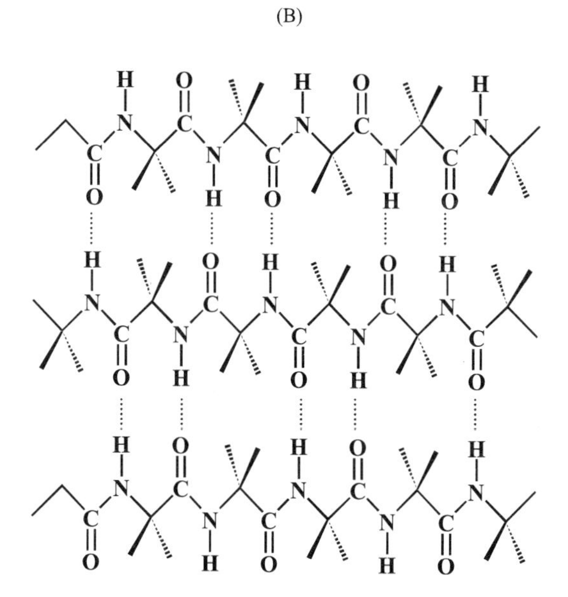

Figure 2.9. Parallel (**A**) and antiparallel (**B**) β-sheets.

interior of the structure, away from the aqueous physiological environment. Polar and ionized side chains are generally oriented toward the outside, where they keep the molecule in solution and prevent aggregation. However, this does not mean that hydrophobic amino acids are never found on the exterior of a protein or that polar and charged amino acids are never buried inside. When on

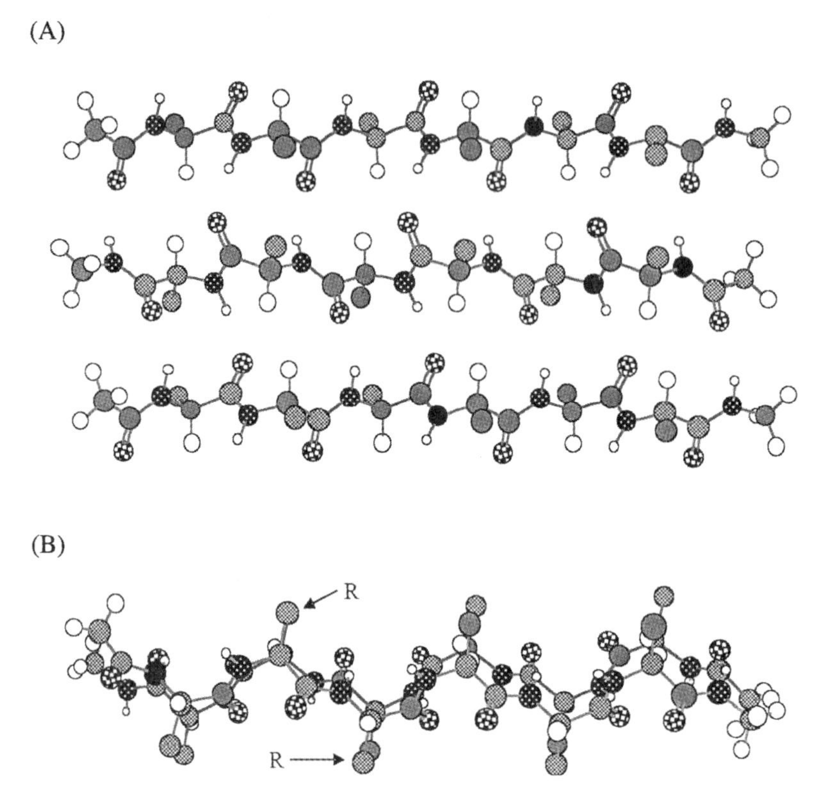

Figure 2.10. β-Sheet of poly(Ala–Gly) as model silk fibroin peptides. (**A**) Top view. (**B**) Side view showing pleated appearance with side chains projecting upward and downward.

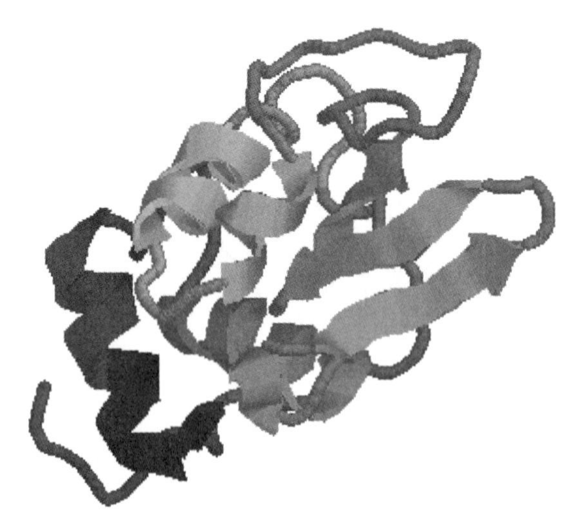

Figure 2.11. Model of the human enzyme lysozyme showing regions of α-helix (spirals), β-strands (arrows), and other structures. Figure created with coordinates available in the Protein Data Bank (accession code 1REX).

the outside, nonpolar residues are frequently dispersed among polar amino acids to minimize adverse effects because of their inability to interact with water. When hydrophobic amino acids are clustered on the outside, it is for a specific purpose, such as for binding other polypeptides (see Section 2.5 on quaternary structure) or substrate molecules. When a charged amino acid is in the interior of the protein, it is also for a specific purpose, such as stabilizing the structure or forming an active binding site.

2.5 QUATERNARY STRUCTURE

Quaternary structure refers to interactions between subunits, or individual polypeptide chains, in multichain proteins. Continuing with the example of a telephone cord, quaternary structure is represented by placing one or more additional cords on or next to the original. The bonds that stabilize the structure are generally the same as those for tertiary structure, except that they occur between amino acids located on different polypeptide chains (Fig. 2.6). The different peptide chains may or may not be identical, each with its own primary, secondary, and tertiary structure. A single subunit or polypeptide chain is called a monomer, two subunits a dimer, three a trimer, four a tetramer, etc. Figure 2.12 shows a hemoglobin monomer and a tetramer.

2.6 IMPORTANCE OF CONFORMATION

The importance of a protein's conformation can be appreciated by considering enzymes, which are proteins that catalyze specific chemical reactions. In these reactions, the enzyme binds a substrate molecule, and consequently, a chemical reaction such as hydrolysis or phosphorylation occurs. Consider the protease trypsin, which is a digestive enzyme that catalyzes the breakdown of proteins into amino acids that can be absorbed in the intestines. The specificity of the reaction results from binding of the substrate in a particular site of the molecule. The structure of the enzyme is organized such that the substrate-binding site has the molecular dimensions and arrangement of functional groups to mediate specific binding. Figure 2.13 shows part of a substrate molecule fitting into the active site of trypsin. Because of the folding of the polypeptide chain, the active site is composed of amino acids from different regions. Based on sequential numbering with respect to trypsin's primary structure, amino acids at positions 57 (histidine), 102 (aspartic acid), and 195 (serine) act together to position the substrate and to cleave a peptide bond. If the secondary or tertiary structure of trypsin was altered, the substrate molecule might not be recognized by the active site. Denaturation of the protein can result from changes in temperature, pH, and ionic strength, which interfere with the various types of intramolecular bonds that stabilize the conformation of the enzyme. For example, changes in

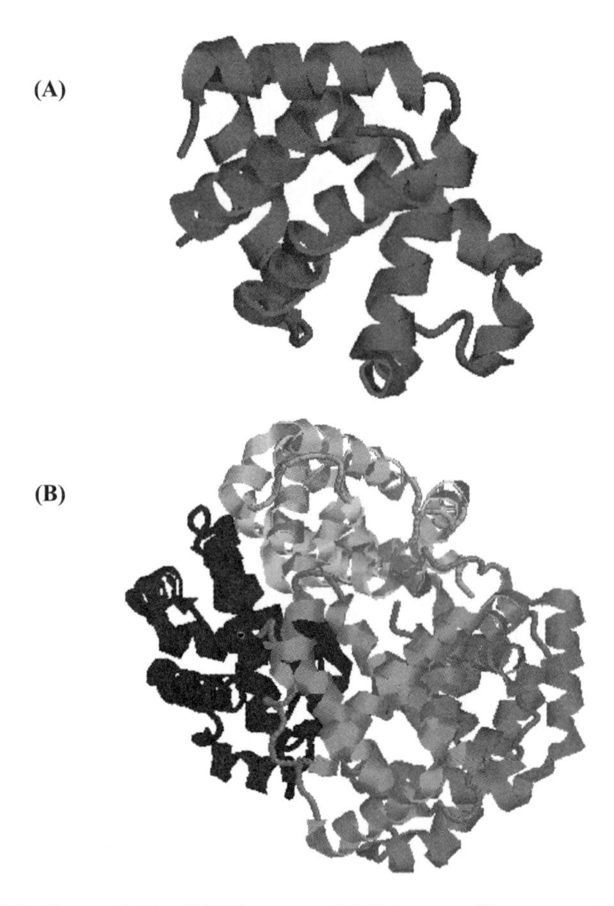

Figure 2.12. Model of hemoglobin. (**A**) Monomer. (**B**) Tetramer. Figures created with coordinates available in the Protein Data Bank (accession codes 1ITH and 1CBL).

pH can alter the ionization of side chains (see Fig. 2.2), which can subsequently interfere with ionic bonding between amino acids. Similar considerations apply to proteins important in mediating interactions between cells and biomaterials at the tissue-implant interface. Furthermore, as discussed in Chapter 3, binding of proteins to surfaces provides additional means for changing the conformation of the biomolecules.

2.7 EXAMPLES

2.7.1 Collagen

Collagen is the most abundant protein found in higher vertebrates. This family of proteins accounts for 25% or more of total body protein and provides an im-

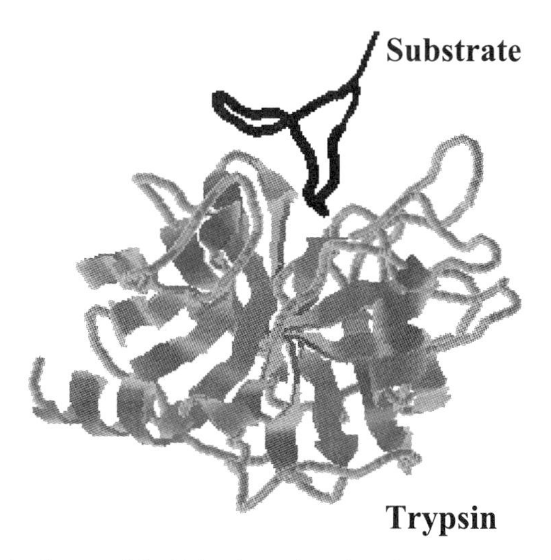

Figure 2.13. Binding of substrate (black) in the active site of trypsin (gray). Only a portion of the substrate molecule (soybean trypsin inhibitor) is shown. A change in the conformation of trypsin would alter the ability of the substrate to fit in the binding site. Figure created with coordinates available in the Protein Data Bank (accession code 1AVW).

portant structural framework for most body structures, including the skeleton, skin, and blood vessels. Nineteen types of collagen have been identified. Type I collagen, found in bone, tendons, ligaments, skin, and many organs, is the most abundant. The fibrillar structure exhibited by the most common collagens is described below.

The primary structure of collagens can be represented by the motif of Gly–Pro–Hyp–Gly–X–Y, where Hyp is hydroxyproline and X and Y can be any amino acids. Glycine constitutes about one-third of the residues, proline about 13%, and hydroxyproline about 10%. Because only small quantities of hydroxyproline are present in other proteins, assays for hydroxyproline are frequently used to determine collagen content. As mentioned in Section 2.3, the cyclic structure of proline causes steric limitations on secondary structure. Consequently, the proline-rich polypeptide chains of collagen form left-handed helices stabilized by hydrogen bonding (Fig. 2.14). Although they are called α-chains, they should not be confused with α-helices. Three α-chains are then wound around each other to form a right-handed procollagen superhelix (Fig. 2.14). The presence of glycine, which has only an H atom as its side group, at every third amino acid allows the three α-chains to pack tightly together; amino acids with bulkier side chains would prevent packing. The triple helix is stabilized by hydrogen bonding between glycine and hydroxyproline and hydroxylysine. The composition of the individual α-chains in the triple helix determines the type of collagen. For example, type I collagen consists of two identical $\alpha 1(I)$-chains and an $\alpha 2(I)$-

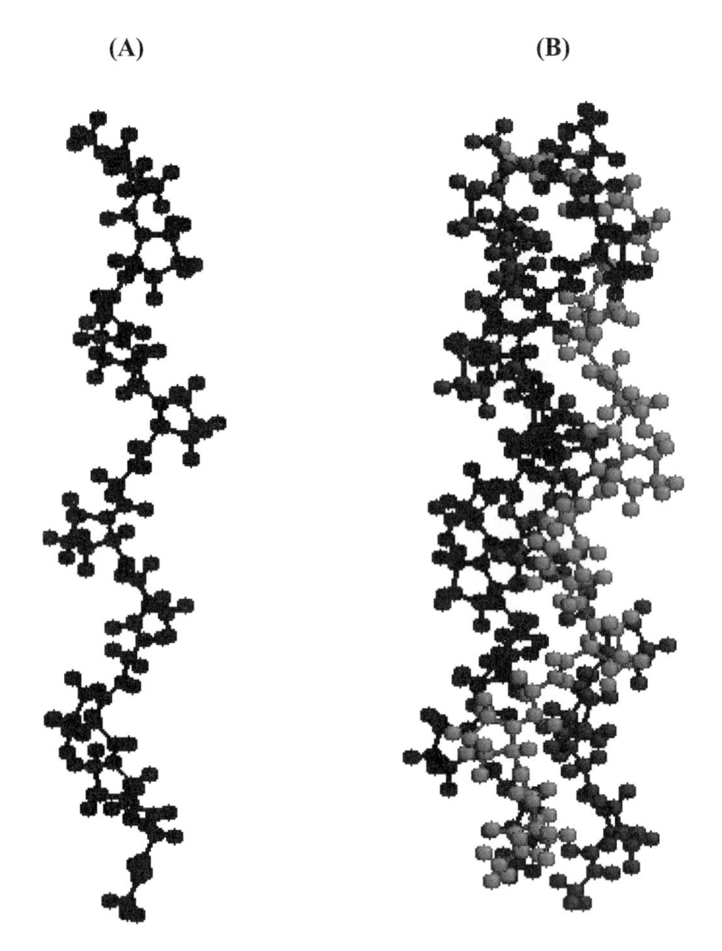

Figure 2.14. Collagen helices. (**A**) Left-handed single chain. (**B**) Right-handed triple helix.

chain. Propeptide regions, believed to play a role in the initial assembly of the triple helix, are present at the ends of procollagen molecules (Fig. 2.15).

After secretion from cells, the propeptides are enzymatically cleaved from the procollagen molecules to form collagen molecules, which are 1.5 nm in diameter and about 300 nm long (Fig. 2.15). Collagen molecules undergo a self-assembly process to form collagen fibrils (10–300 nm in diameter). Assembly results in a "quarter-stagger" array of collagen molecules, in which overlapping rows of molecules are staggered by about one-quarter of the length of an individual molecule (Fig. 2.15). This arrangement gives collagen fibrils characteristic striations every 67 nm when viewed by transmission electron microscopy. Fibrils can aggregate to form collagen fibers, which can be several micrometers in diameter. Formation of enzyme-catalyzed covalent cross-links among lysine residues, both within and between collagen molecules, provides further stabilization of collagen structure (Fig. 2.16). The highly organized structure of collagen fibrils and fibers gives them high uniaxial tensile strength.

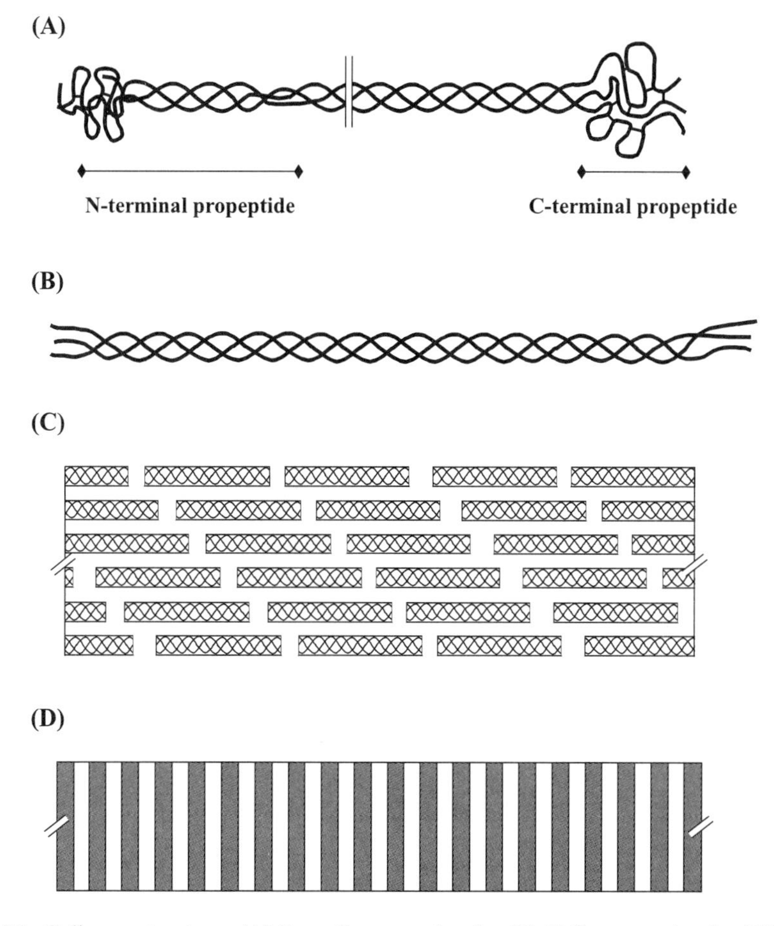

Figure 2.15. Collagen structure. (**A**) Procollagen molecule. (**B**) Collagen molecule. (**C**) Quarter-stagger array of collagen molecules. (**D**) Crossbanding of collagen fibrils resulting from the quarter-stagger structure.

2.7.2 Elastin

Elastin is an important component of tissues, such as skin, blood vessels, ligaments, and lungs, that require elasticity to function. Similar to collagen, elastin is rich in glycine and proline, but these amino acids are not found in a Gly–

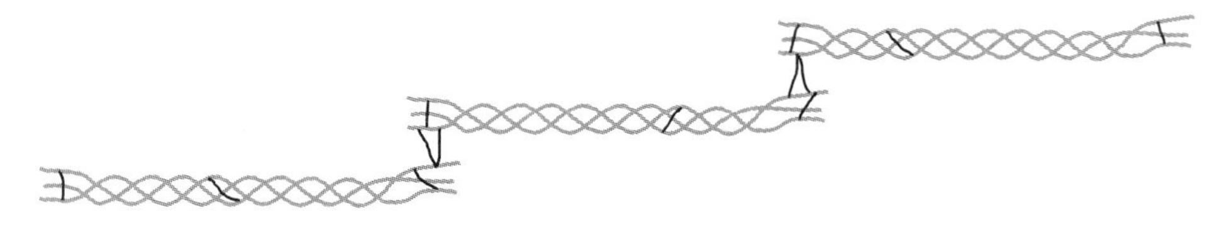

Figure 2.16. Representation of intra- and intermolecular cross-links (dark lines) that stabilize the structure of collagen.

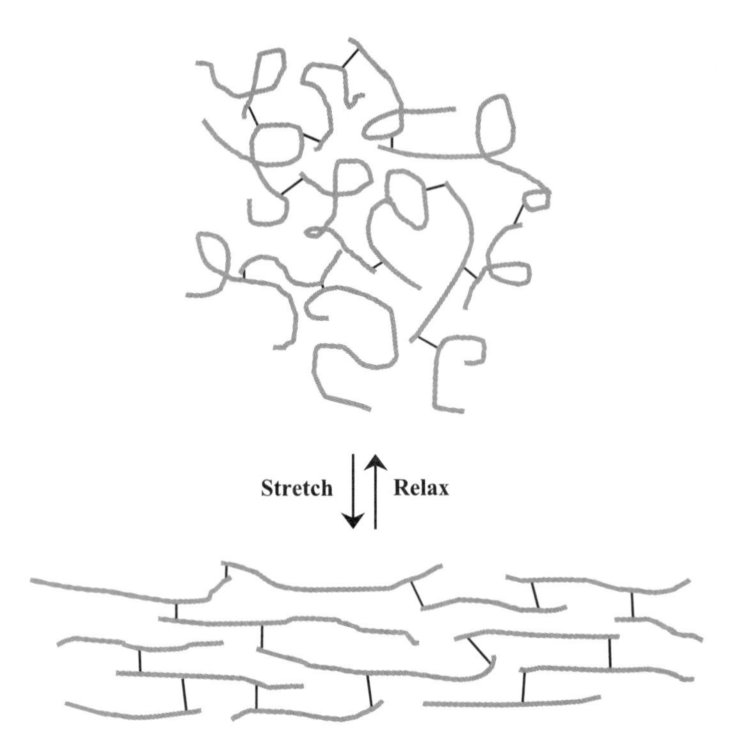

Figure 2.17. Model for the structure of elastin. [Adapted from Alberts, B., Bray, D., Lewis, J., Raff, M., Roberts, K., and Watson, J.D, *Molecular Biology of the Cell*, 3rd Edition, Garland Publishing, New York, 1994.]

Pro–X repeating sequence, and little of the proline is hydroxylated. Hydrophobic amino acids, including alanine, valine, and isoleucine, make up about 50% of the composition of elastin. Elastin chains are composed of alternating segments of hydrophobic amino acids and segments rich in alanine and lysine. Because of the insolubility of mature elastin, the higher levels of structure are still not clear. One model of elastin structure is shown in Figure 2.17. Tropoelastin molecules, analogous to collagen molecules, are secreted from the cell and assemble into elastic filaments, fibrils, fibers, or sheets. In contrast to collagen, rather than having a regular secondary structure, elastin exhibits an unordered coiled structure. The loosely organized assembly is then stabilized by formation of covalent cross-links between lysine residues, as occurs in collagen. In elastin, however, the cross-links can involve not only two lysines, but four as well (Fig. 2.18). In the relaxed state, the molecules exist as a covalently cross-linked network of loose and unorganized polypeptide chains. As elastin fibers are stretched, the chains become aligned, and the cross-links between lysine residues constrain further extension once the chains are straight (Fig. 2.17). When unloaded, the chains return to their original coiled states because of intramolecular hydrophobic interactions. In this way, tissues containing elastin behave similar to rubber bands and are able to deform and recover without damage.

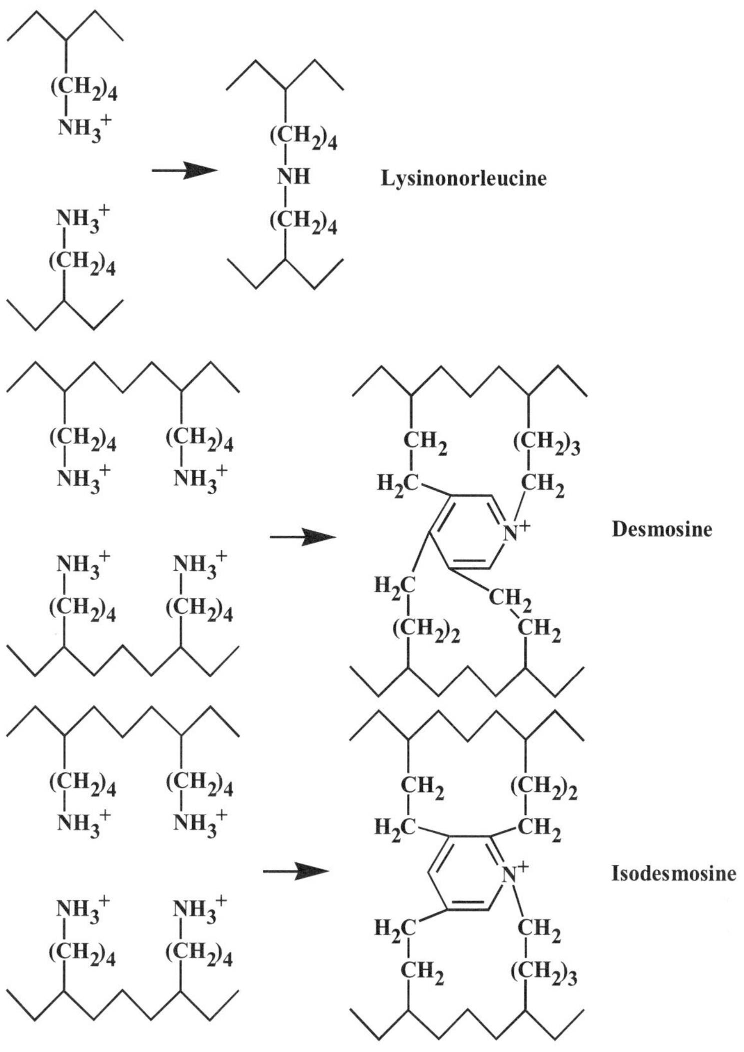

Figure 2.18. Cross-links that stabilize the structure of elastin.

2.7.3 Fibronectin

The extracellular matrix contains noncollagenous adhesive proteins that play important roles in organizing the matrix and in enabling cells to attach to it. The most studied of these proteins is fibronectin. Fibronectin is composed of two subunits, each with a molecular weight of about 220,000 Da, joined by two disulfide bonds near their carboxyl termini (Fig. 2.19). The subunits have a modular architecture constructed of repeats of three types of structures, classified as type I, II, and III repeats. The segments, 40–90 amino acids long, con-

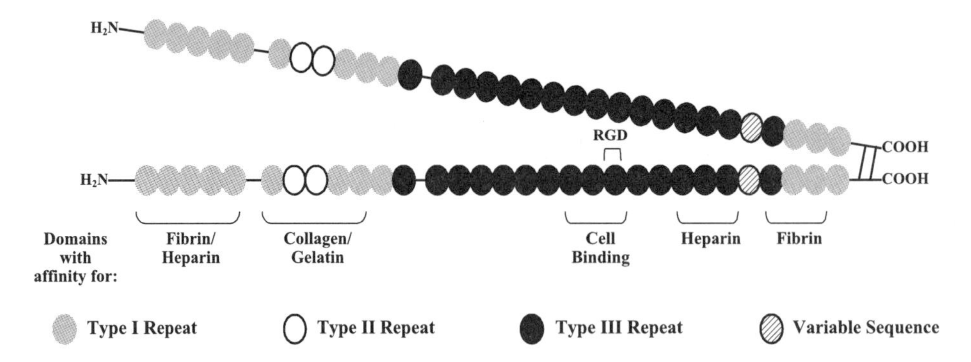

Figure 2.19. Schematic representation of fibronectin showing its modular architecture and functionality of various domains. [Adapted from Magnusson, M.K. and Mosher, D.F. (1998). Fibronectin: Structure, assembly, and cardiovascular implications. *Arterioscler. Thromb. Vasc. Biol.* 18: 1363–1370.]

tain β-strands and β-sheets (Fig. 2.20). The modules are serially repeated with linker peptides between segments. Multifunctionality of fibronectin is provided by domains within the subunits having affinity for various molecules, such as cell surface receptors, collagen, and heparin (Fig. 2.19). Of particular importance is the cell-binding tripeptide of Arg–Gly–Asp (RGD), which in synergy with adjacent peptide sequences binds to specific cell surface receptors and results in cell adhesion (Figs. 2.19 and 2.20).

2.7.4 Fibrinogen

Fibrinogen presents a more complex structure. This is a large plasma protein that plays a critical role in blood clotting as well as several other physiological and pathological processes. Although it has become apparent that several variants of fibrinogen circulate in blood, a common form is a 340-kDa rod-shaped molecule, with a length of 40–50 nm and a width of 4–10 nm. This protein is composed of two identical subunits, each consisting of three different polypeptide chains, which are denoted Aα, Bβ, and γ (Fig. 2.21). The six chains are bound together in a central "disulfide knot." Fibrinogen is described as having three structural regions: two globular terminal D domains linked by triple helical segments to a central E domain, which contains the disulfide knot. At least 18 distinct domains have been identified in fibrinogen. Calcium ions, which are essential for the functional integrity of fibrinogen, interact with the protein via low-affinity and high-affinity binding sites. Multiple cell attachment sites, two of which contain the RGD sequence, have been identified. Various cells involved in blood clotting and tissue repair can bind to fibrinogen. Other plasma proteins can bind to fibrinogen through both noncovalent and covalent interactions. For

(A)

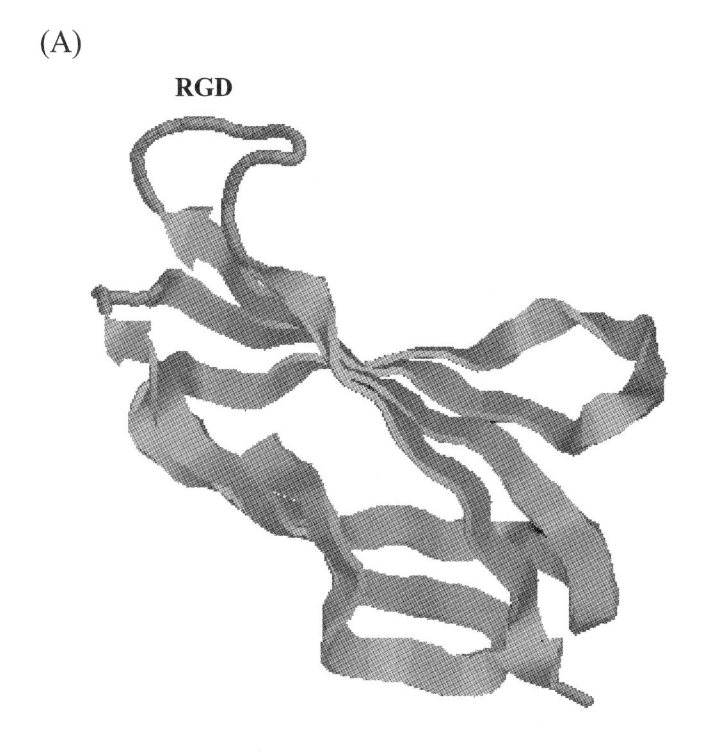

(B)

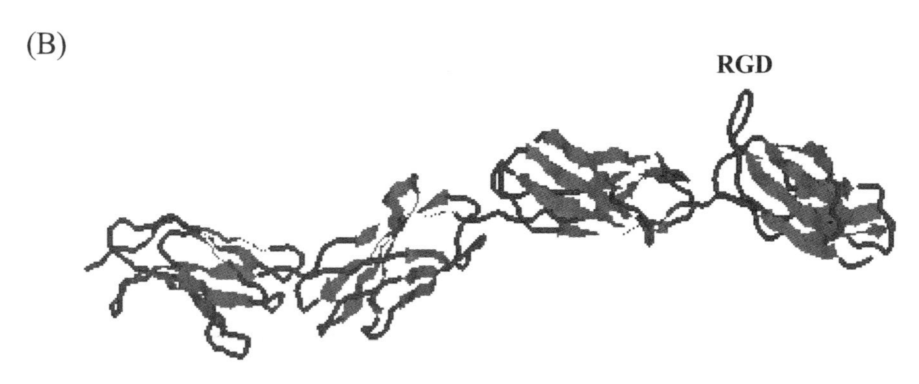

Figure 2.20. Structure of fibronectin. (**A**) The tenth type III repeat containing the RGD cell-binding sequence. (**B**) The seventh through tenth type III repeats. Figures created with coordinates available in the Protein Data Bank (accession codes 1FNA and 1FNF).

example, fibronectin can be covalently linked to fibrinogen through an enzymatically catalyzed reaction between lysine residues on fibrinogen and glutamine residues on fibronectin. Sites for proteolytic cleavage are also present. As discussed in other chapters, these cleavage products play important roles in coagulation and inflammation.

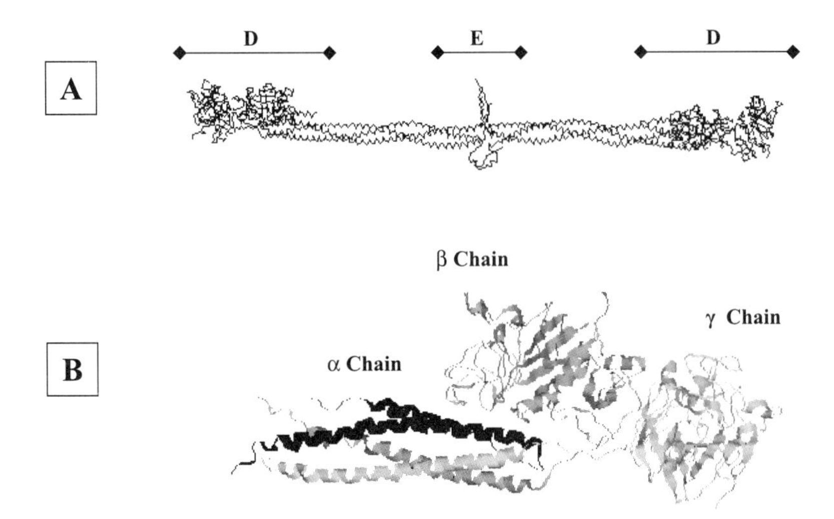

Figure 2.21. Structure of fibrinogen. (**A**) The entire rod-shaped molecule, showing the two terminal D regions and the central E region containing the disulfide knot. (**B**) One end of fibrinogen, showing the globular termini of the Bβ- and γ-chains in the D region as well as the triple-helical nature of the segments leading to the E region. Figures created with coordinates available in the Protein Data Bank (accession codes 1EI3 and 1FZA).

2.8 SUMMARY

- Proteins are chains of amino acids, whose side groups can be nonpolar, polar, or charged.
- Interactions between amino acids are governed by hydrogen, ionic, and covalent bonding and hydrophobic interactions.
- The primary structure of a protein is its linear sequence of amino acids.
- Secondary structure refers to localized coiling and bending of the polypeptide chain.
- Tertiary structure refers to the overall folding of an entire protein subunit.
- Quaternary structure refers to association of multiple subunits, which may be identical or each may have its own primary, secondary, and tertiary structure.
- The properties and functions of proteins, such as collagen, elastin, fibronectin, and fibrinogen, are directly related to the intramolecular interactions resulting from their primary structures.

2.9 BIBLIOGRAPHY/SUGGESTED READING

Branden, C. and Tooze, J. *Introduction to protein structure*, 2nd Edition, Garland Publishing, New York, 1999.

Voet, D. and Voet, J. *Biochemistry*, 2nd Edition, John Wiley & Sons, New York, 1995.

2.10 QUIZ QUESTIONS

1. Draw the structure of an amino acid.
2. What is the chemical process by which a polypeptide is formed?
3. Describe two common secondary structures observed in proteins.
4. What forces stabilize protein structure? Describe how each type can be disrupted.
5. Why does type I collagen have a crossbanded appearance when visualized by transmission electron microscopy?
6. What gives elastin its elasticity?
7. What does "RGD" refer to? Explain its significance.

2.11 STUDY QUESTIONS/DISCOVERY ACTIVITIES

1. Many congenital conditions can be attributed to genetic mutations that result in substitutions of amino acids. Discuss the effects on a protein's structure of a hypothetical substitution of tryptophan for glutamic acid.
2. What is meant by protein conformation? Relate changes in conformation to potential events at the tissue-implant interface.
3. Compare and contrast the structures of collagen and elastin. How do these structures relate to their properties?
4. Detailed molecular structures of numerous proteins have been published, and the number continues to increase. Search the Protein Data Bank (http://www.rcsb.org/pdb) for the structure of a protein important to tissue formation and repair, such as bone morphogenetic protein 2, vascular endothelial cell growth factor, or nerve growth factor. Using a plug-in, such as Chime, or a program, such as Rasmol, view the growth factor's structure. Rotate the molecule. Describe its structure using the terms presented in this chapter.

3

Protein-Surface Interactions

3.1 INTRODUCTION

The behavior of proteins at surfaces plays a vital role in determining the nature of the tissue-implant interface. Adsorbed proteins affect blood coagulation, complement activation, and bacterial and cell adhesion. Furthermore, adsorbed proteins can influence biomaterial surface properties and degradation. This chapter presents the basic principles of protein adsorption in terms of protein and surface characteristics that affect the behavior of proteins at solid surfaces.

3.2 IMPORTANT PROTEIN AND SURFACE PROPERTIES

The properties of both the protein and the surface with which the biomolecule is interacting influence interfacial behavior (Fig. 3.1). Tables 3.1 and Table 3.2 list important protein and surface properties, respectively.

3.2.1 Protein Properties

The properties of proteins that influence surface activity are related to the primary structure of the protein, meaning that the sequence of amino acids affects protein-surface interactions. Larger molecules are likely to interact with surfaces because they are able to contact the surface at more sites (Fig. 3.2). For example, an albumin molecule (67 kDa) forms about 77 contacts with a silica substrate, and fibrinogen (340 kDa) forms about 703 contacts per molecule. Size, however, is not the sole determinant, because hemoglobin (65 kDa) exhibits greater surface activity than the much larger fibrinogen.

Because of their hydrophilicity, charged amino acids are generally located on the outside of proteins and are readily available to interact with surfaces. Consequently, the charge, as well as the distribution of charge on the protein surface, can greatly influence protein adsorption. As with size, however, charge is not the only determinant. Interestingly, proteins often show greater surface activity near their isoelectric point (the pH at which the molecule exhibits zero

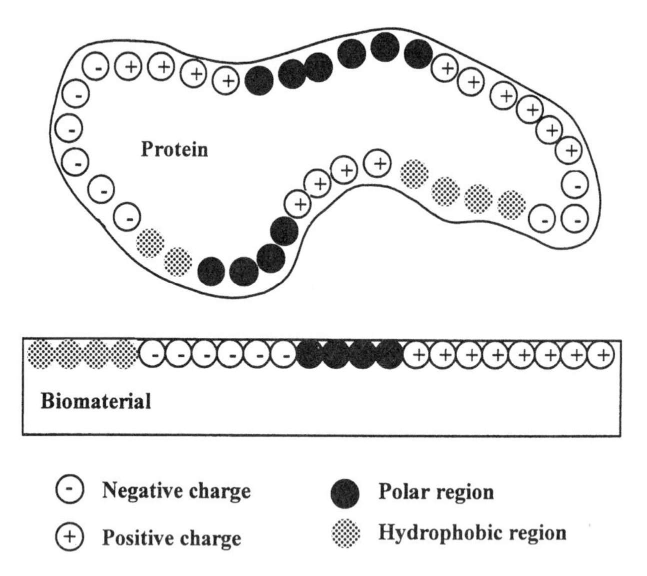

Figure 3.1. Cartoon illustrating the importance of both molecular and substrate properties in determining protein-surface interactions.

TABLE 3.1. Properties of Proteins That Affect Their Interaction With Surfaces

Property	Effect
Size	Larger molecules can have more sites of contact with the surface
Charge	Molecules near their isoelectric point generally adsorb more readily
Structure	
Stability	Less stable proteins, such as those with less intramolecular cross-linking, can unfold to a greater extent and form more contact points with the surface
Unfolding rate	Molecules that rapidly unfold can form contacts with the surface more quickly

TABLE 3.2. Properties of Surfaces That Affect Their Interaction With Proteins

Feature	Effect
Topography	Greater texture exposes more surface area for interaction with proteins
Composition	Chemical makeup of a surface will determine the types of intermolecular forces governing interaction with proteins
Hydrophobicity	Hydrophobic surfaces tend to bind more protein
Heterogeneity	Nonuniformity of surface characteristics results in domains that can interact differently with proteins
Potential	Surface potential will influence the distribution of ions in solution and interaction with proteins

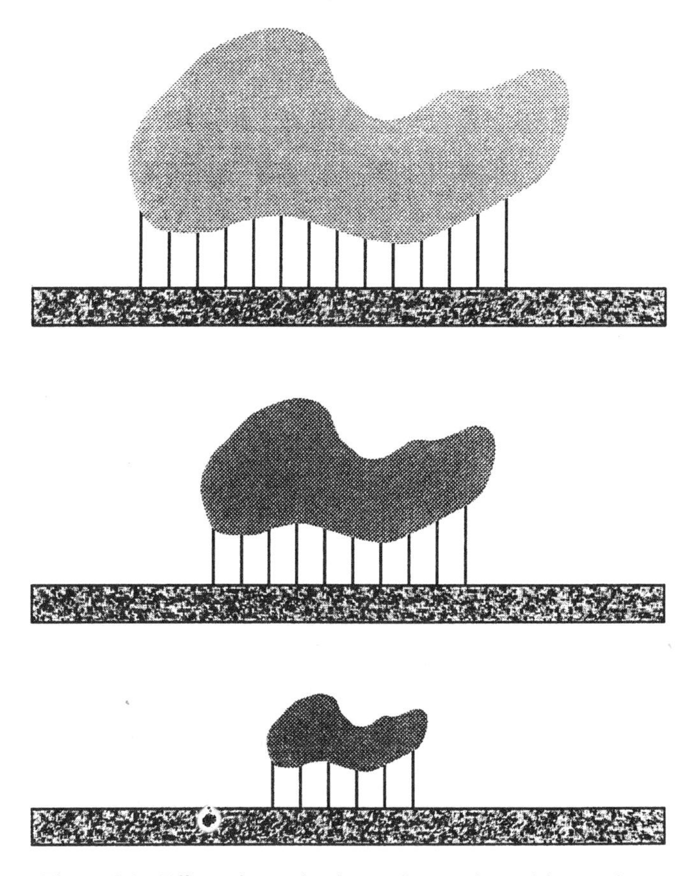

Figure 3.2. Effect of protein size on interaction with a surface.

charge, denoted pI). Although at first thought this might seem odd, two effects can explain the observation. First, consider that protein molecules do not interact with the surface in isolation. For example, 1 ml of a solution containing 1 μg of a 50-kDa protein will have approximately 10^{13} molecules. With so many molecules in solution, not only can they interact with the surface, but the molecules can also interact with each other (lateral interactions). At the isoelectric point, reduced electrostatic repulsion between uncharged adsorbing molecules can allow more protein to bind. A second explanation relates to alterations in protein structure because of changes in the charge of amino acids (see Chapter 2). If the conformation is altered, different amino acids could be exposed on the surface of the protein, which could consequently change the way the molecule binds to the substrate.

Properties related to unfolding of the protein also affect adsorption. Unfolding of a protein is likely to expose more sites (points) for protein-surface contact (Fig. 3.3). Therefore, factors related to a greater extent or rate of unfolding can result in greater surface activity. Less stable proteins or those with less intramolecular cross-linking are likely to unfold more or faster. For example,

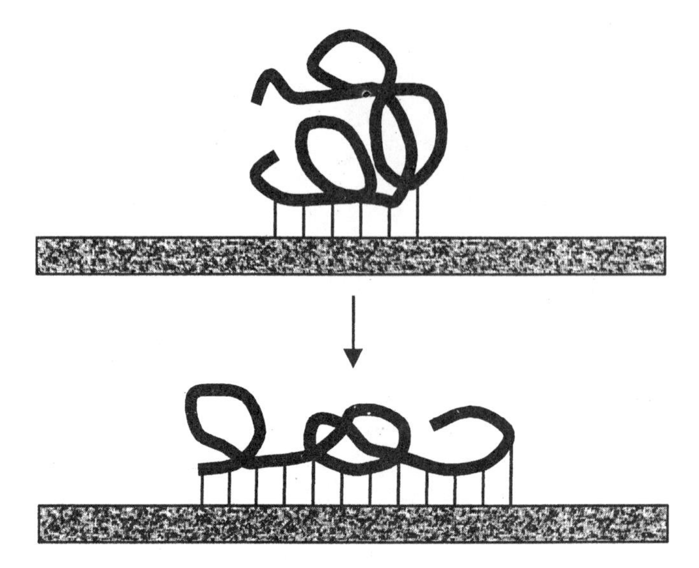

Figure 3.3. Effect of protein unfolding on interaction with a surface.

substitution of hydrophobic valine for glutamic acid in hemoglobin makes hemoglobin S (see Chapter 2, Section 2.2) less stable. The destabilized protein is consequently less soluble and results in fibrous precipitates that distort red blood cells.

The amphipathic nature of proteins, with their polar, nonpolar, and charged amino acids, also contributes to surface activity (Fig. 3.1). Although hydrophilic polar and charged amino acids are generally located on the exterior of the molecule and hydrophobic residues on the interior, this is not absolute. Thus hydrophobic amino acids might be available on the protein surface for interaction with substrates. Additionally, unfolding of proteins can expose hydrophobic regions and allow interaction with the surface.

3.2.2 Surface Properties

The properties of biomaterial surfaces that influence interaction with proteins are similar to those for proteins. In discussing surface properties, they are frequently grouped in three categories: geometric, chemical, and electrical. Substrates with more topographical features will expose more surface area for possible interaction with proteins. For example, surfaces with grooves or pores have greater surface area compared with smooth surfaces. Other surface features, such as machine marks introduced during processing, provide additional sites for protein interaction.

The surface chemical composition will determine which functional species are available for interaction with biomolecules. The oxidized (passivated) surface of a metallic biomaterial exposes metal and oxygen ions. Similarly, ceramic, and some glass, surfaces comprise metal and nonmetal ions. A variety of functional species, such as amino, carbonyl, carboxyl, and aromatic groups, can be present on the surface of polymeric biomaterials. Depending on which species are exposed, biomolecules (or even particular regions of the molecule) may have different affinities for various surfaces. For example, hydrophobic surfaces tend to bind more protein as well as binding it more tenaciously.

On a microscopic scale, biomaterial surfaces can be inhomogeneous. Patches, or domains, of different functionality can exist on biomaterial surfaces, and these patches can interact differently with biomolecules. For example, many metallic biomaterials contain at least two different phases, such as the α- and β-phases in Ti-6Al-4V. Not only can the different phases behave differently when interacting with biomolecules, but grain boundaries behave differently than do grain interiors. In polymers, segregation resulting from folding of macromolecular chains can give microstructural domains. Depending on the chemical species present within the various domains, proteins will have different affinities for the patches.

The surface potential influences the structure and composition of the electrolyte solution adjacent to the biomaterial. Counterions are attracted to the surface, and normally isotropically distributed water molecules become ordered. The combined effects of water ions, molecules, and net surface potential will determine whether interaction with biomolecules is enhanced or hindered.

3.3 ADSORPTION AND DESORPTION

Adsorption is the process whereby molecules adhere to solid surfaces. Protein-surface interactions result in high local concentrations of the protein, reaching concentrations up to 1,000 times higher than in the bulk solution. As discussed in other chapters, this accumulation of protein, and especially accumulation of certain proteins, on biomaterial surfaces plays a critical role in determining the fate of the tissue-implant interface.

In addition to the protein and surface properties described above, adsorption also depends on the availability of molecules for interaction with the substrate. Molecules can be brought to the surface by one or more of four major transport mechanisms: 1) diffusion, 2) thermal convection, 3) flow, and 4) coupled transport, such as the combination of convection and diffusion. Variables such as concentration, velocity, and molecular size are important in determining the arrival of protein molecules at a surface. As an example, consider the effects of diffusion. Simple diffusion is described by the following equation:

$$\frac{\delta C}{\delta t} = D\frac{\delta^2 C}{\delta x^2} \tag{3.1}$$

where C is concentration, D is the diffusion coefficient, and x is distance. At short times and under conditions in which the rate of adsorption equals the rate of diffusion

$$\frac{dn}{dt} = C_0 \left(\frac{D}{\pi t}\right)^{1/2} \tag{3.2}$$

where n is the surface concentration of protein, C_0 is the bulk concentration of protein, and t is time. *Equation 3.2* shows that a higher bulk concentration and/or higher diffusion coefficient (which is inversely related to molecular size) result in a larger number of molecules arriving at the surface. Under conditions in which convection is also present, resulting in convective diffusion, the treatment becomes more complex and depends on the geometry of the interface. For flow in a thin channel:

$$\frac{\delta C}{\delta t} + V(y)\frac{\delta C}{\delta x} = D\frac{\delta^2 C}{\delta y^2} \tag{3.3}$$

where

$$V(y) = \gamma y\left(1 - \frac{y}{b}\right) \tag{3.4}$$

and V is the velocity of flow, x is the distance down the channel, y is the location within the height of the channel, γ is the wall shear rate, and b is the height of the channel. After applying the pertinent boundary conditions, *Equation 3.4* must be solved with numerical methods.

Once present at the surface, protein molecules can interact with the substrate via intermolecular forces, such as ionic bonding, hydrophobic interactions, and charge-transfer interactions. In contrast to its importance in stabilizing protein structure, hydrogen bonding does not play a major role in protein-surface interactions. Because water is good at forming hydrogen bonds, it is as likely to form hydrogen bonds with a surface as would amino acids in the protein molecule. Exactly which intermolecular forces govern protein-surface interaction will depend on the particular protein and surface (Fig. 3.1).

Even with a solution containing a single type of protein, the layer of adsorbed protein is likely to be heterogeneous. As molecules adsorb to a clean surface, there are few limitations on their interaction with the substrate, and each molecule can form many contacts with the surface (Fig. 3.4). As the surface becomes occupied, however, less surface is available for adsorption of subsequent protein molecules. Consequently, molecules in different orientations might be able to bind to the surface, even though fewer protein-substrate contacts are made (Fig. 3.4). Different orientations can also allow the protein to avoid or minimize repulsive interactions with previously bound biomolecules. In addition to reasons of available contact area, proteins can exist on the surface in different orienta-

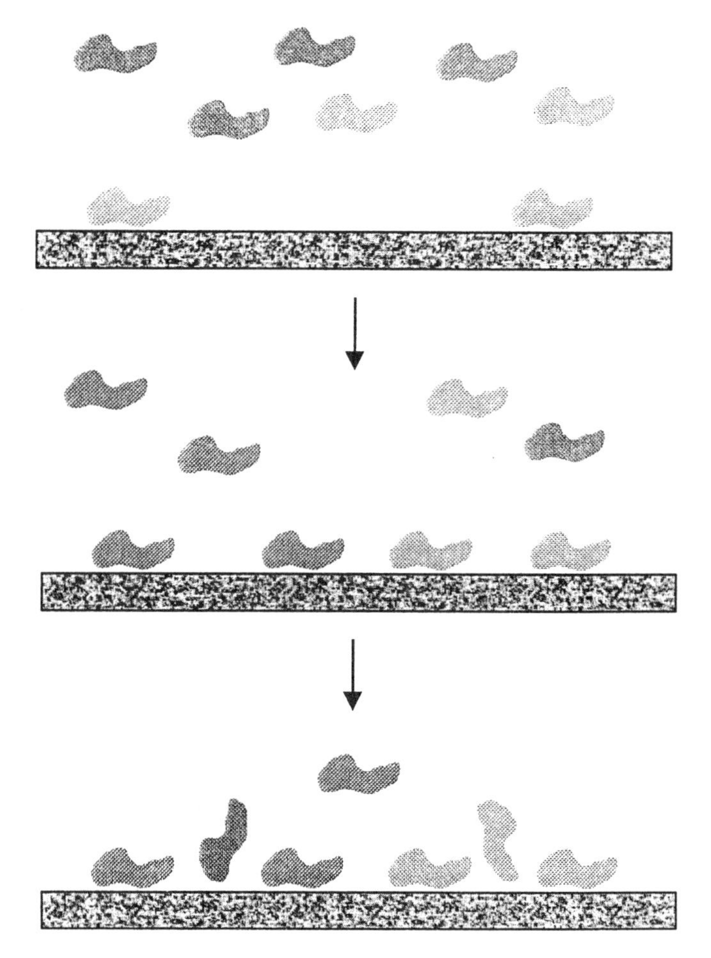

Figure 3.4. Effect of surface occupancy on adsorption of subsequent protein molecules. [Adapted from Horbett, T.A., and Brash, J.L. Proteins at interfaces: Current issues and future prospects, *Proteins at Interfaces: Physiochemical and Biochemical Studies*, J.L. Brash and T.A. Horbett (eds.), American Chemical Society, Washington, D.C., 1987, pp. 1–33]

tional states because of the heterogeneity of both the protein molecule and surface (Fig. 3.5); different orientations may be needed to bring complementary functionalities on the surface and protein into close proximity. For example, amphipathic proteins, with their polar, nonpolar, and charged amino acids, can interact differently with the biomaterial's various microstructural features, which have distinct structural and chemical properties.

The different orientations of adsorbed protein molecules not only affect the amount of protein bound to the surface but also have functional significance. Consider an enzyme or an adhesive protein, such as fibronectin, adsorbing to a biomaterial. Depending on the orientation of the molecules, the active site needed for catalytic activity of the enzyme could be inaccessible, either because it is interacting with the surface or because access to the active site is prevented

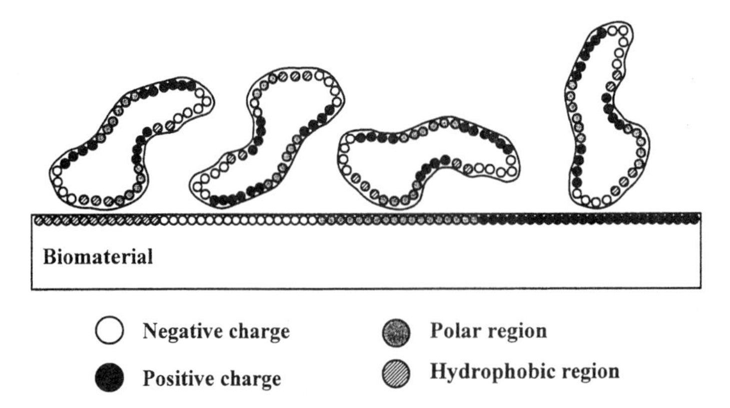

Figure 3.5. Different orientations of adsorbed protein molecules resulting from heterogeneity of both the protein and the surface.

by adjacent molecules (steric hindrance). Similarly, if the RGD-containing domains of fibronectin (Fig. 2.19) are not available for interaction with cells, the protein may not be able to support cell attachment.

Desorption is the reverse of adsorption; molecules previously bound to a surface detach and return to the bulk phase. For desorption to occur, all contacts between protein and surface must be simultaneously broken (Fig. 3.6). Although well characterized for small molecules such as gases, desorption of proteins is slow or nonexistent. Unless dramatic changes are made in the interfacial environment, such as increased ionic strength, lowered pH, and use of chaotropic agents or detergents, protein adsorption is largely irreversible because of the requirement of simultaneous dissociation of all interactions between molecule and surface. The difficulty or improbability of simultaneous disruption

Figure 3.6. Desorption of a protein requires simultaneous breakage of all bonds with a surface.

of all contacts is increased further by large proteins, which can form a greater number of bonds with the surface. Consider the example of fibrinogen adsorption mentioned in Section 3.2.1. For a molecule of fibrinogen to desorb, all 703 contacts with the surface must be broken at the same time. As discussed in Section 3.5, however, adsorbed proteins can be replaced by molecules of the same or a different type of protein.

3.4 CONFORMATIONAL CHANGES

Protein molecules must not be thought of as rigid structures. As discussed in Chapter 2, proteins are flexible chains that have been coiled, folded, and bent to assume a particular conformation (three-dimensional structure). Changes in the microenvironment of the proteins, such as pH and ionic strength, can alter the conformation of the molecule. Likewise, proteins experience structural alterations during interaction with solid surfaces. Their conformation may be changed, but adsorbed proteins generally retain at least some of their biological activity. For example, adsorbed enzymes retain their catalytic properties, and adsorbed antibodies retain their ability to bind antigen, although the level of activity may be diminished.

Two modes of conformational change can occur. First, protein molecules can undergo time-dependent molecular spreading (Fig. 3.7). Initially, the molecule may contact a minimal number of binding sites on the surface by interaction of amino acids on the exterior of the protein. As the length of time the molecule resides on the surface (residence time) increases, the protein may unfold, exposing interior functional groups for interaction with additional binding sites. Overall, this results in a time-dependent increase in the number of contact points between protein and surface. Consequently, desorption becomes less likely as the residence time increases because of the larger number of contacts formed.

Second, altered conformation can result from changes in the bulk solution concentration (Fig. 3.8). At low concentration, abundant surface area is available for each protein molecule. Without near neighbors, molecules can spread to form multiple contacts with the surface. At high bulk concentrations, the amount of surface per molecule decreases and less unfolding can occur, because of adsorbate-adsorbate interactions. Consequently, more protein may be present on the surface, but with each molecule having fewer contacts.

3.5 MULTICOMPONENT SOLUTIONS

Although much of the understanding of protein-surface interactions described above has come from the study of single-protein solutions, adsorption from multicomponent solutions is most relevant to the tissue-implant interface. Body fluids, including blood, tears, and saliva, contain numerous types of biomole-

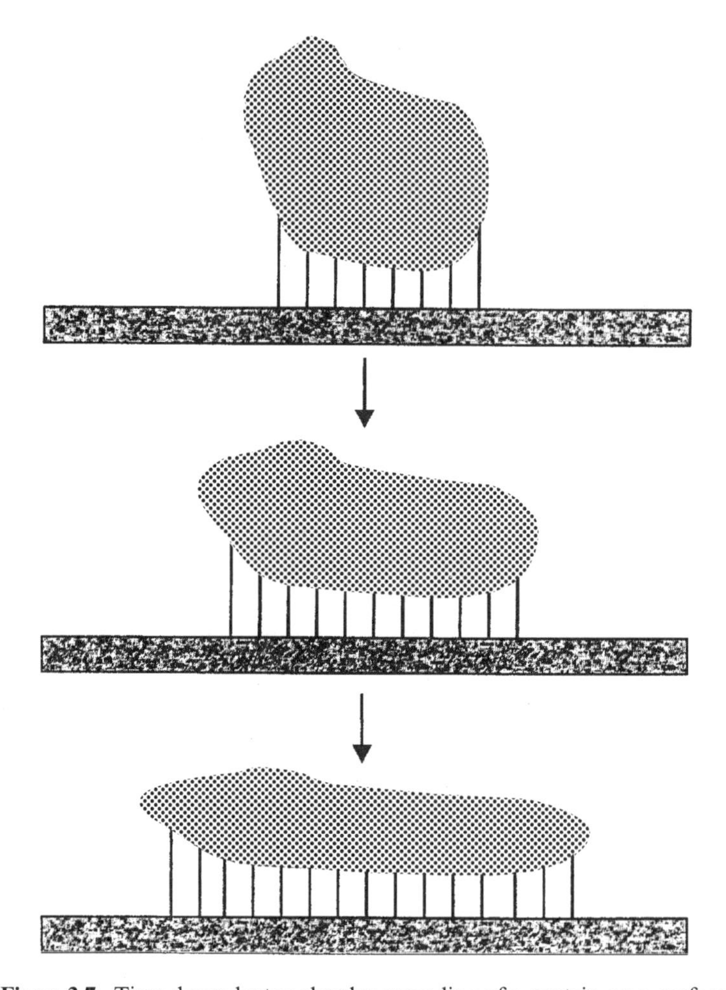

Figure 3.7. Time-dependent molecular spreading of a protein on a surface.

cules. For example, blood contains more than 150 proteins, not to mention lipids, carbohydrates, hormones, etc. Of ultimate importance is knowing which molecules accumulate on a biomaterial surface and how this relates to the bulk composition of the solution.

When a surface is exposed to a multicomponent solution, certain molecules will be preferentially deposited from the bulk. Furthermore, time-dependent changes in the composition of the adsorbed layer can occur, until a pseudo-steady state is reached. Variables related to both surface activity and availability of biomolecules at the surface contribute to determining the profile of molecules on the surface. Thus the affinity (e.g., size, charge, and conformational stability) and kinetic factors (e.g., concentration and size) described above are important. Because surfaces present a finite amount of area for binding protein, molecules

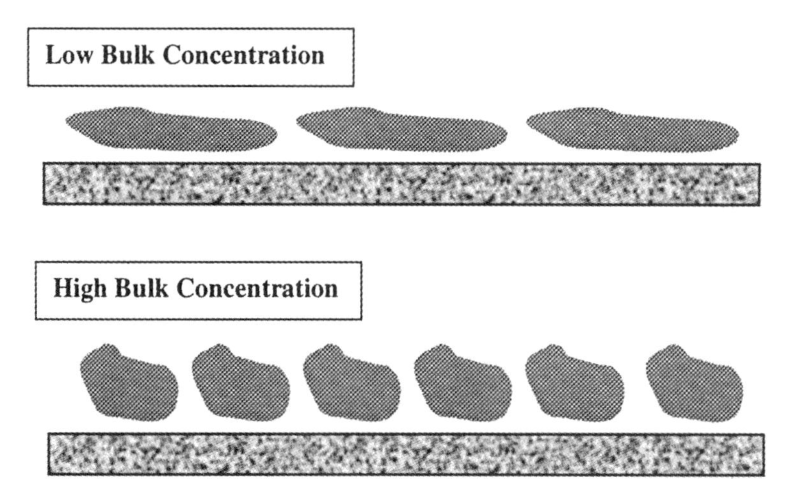

Figure 3.8. Conformational changes depending on the concentration of a protein in the bulk solution.

approaching the surface compete for binding sites, and protein-protein interactions as well as protein-surface interactions are important. In single-protein solutions intermolecular repulsive interactions dominate, but with multicomponent systems attraction between molecules can occur

Considering a simple diffusion-limited situation at the interface, *Eq. 3.2* indicates that molecules present in the bulk solution at high concentration and/or proteins with small size (large diffusion coefficient) will arrive quickly. Although their affinity for the surface may not be optimal, adsorption, even if just temporary, is likely because of their proximity to a "bare" surface with abundant binding sites. With time, molecules having greater affinity for the surface, but with a slower rate of arrival because of lower concentration and/or larger size, approach. The surface, however, may already be occupied by a monolayer of protein. In this case, the only way new molecules can bind to the surface is if previously adsorbed molecules detach. As stated in Section 3.3, pure desorption is rarely observed. Adsorbed molecules can be exchanged, nonetheless. Exchange results from competition for binding sites between the already adsorbed protein and molecules arriving from the bulk solution (Fig. 3.9). As bonds between the adsorbed molecule and the surface are periodically broken, new protein molecules can occupy the binding sites. The first molecule is released from the surface when all of its contacts with the substrate become occupied by the new molecule. Exchange proceeds until the surface is populated with proteins having strong interaction with the substrate. This hierarchical series of collision, adsorption, and exchange processes has been termed "the Vroman effect" (Fig. 3.10).

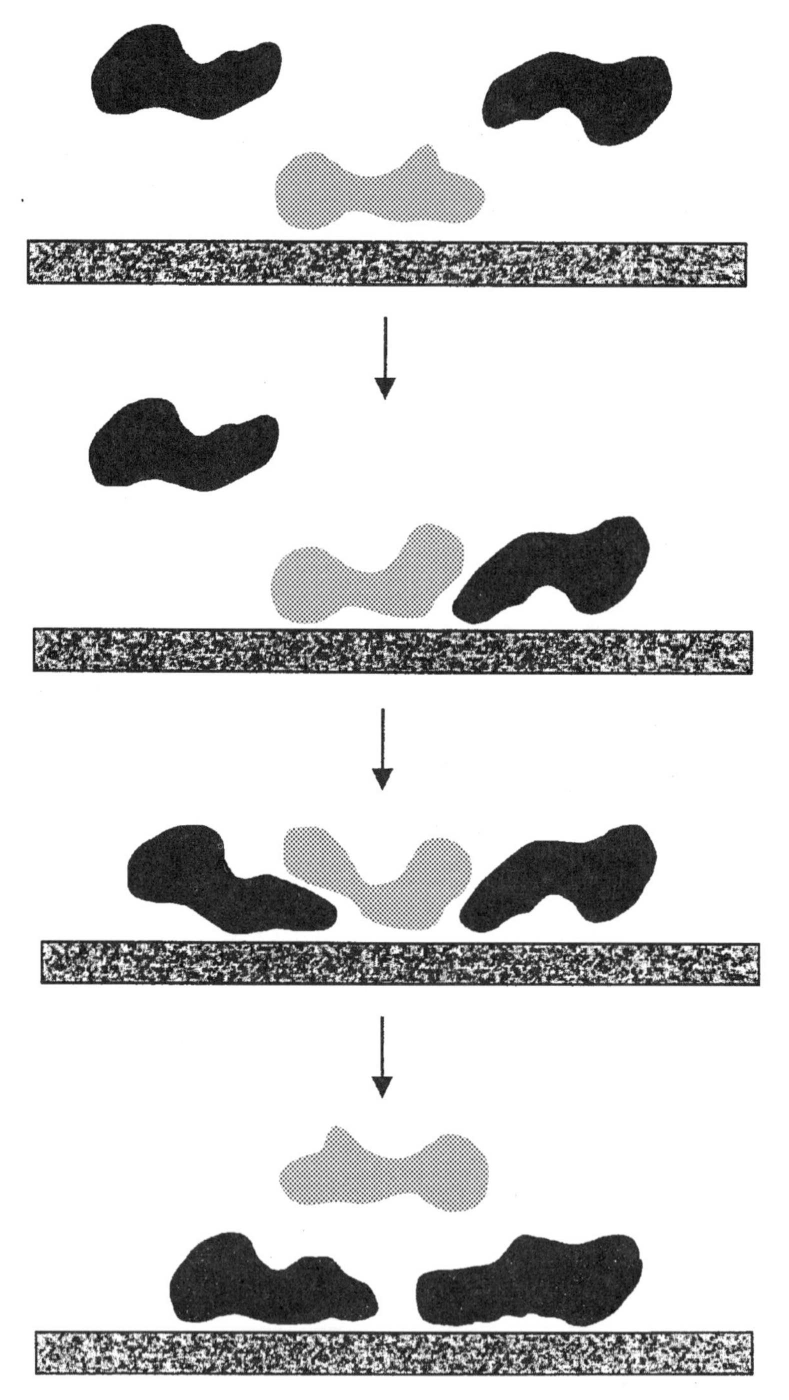

Figure 3.9. Exchange of an adsorbed protein for a different protein. [Adapted from Andrade, J.D., Principles of protein adsorption, *Surface and Interfacial Aspects of Biomedical Polymers*, Vol. 2, J.D. Andrade (ed.), Plenum Press, New York, 1985, pp. 1–80]

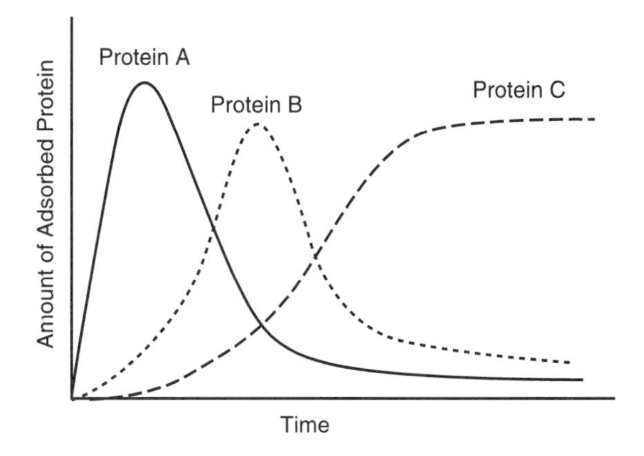

Figure 3.10. Schematic representation of sequential protein exchange on a surface (the Vroman effect).

3.5.1 Example—Blood-Surface Interactions

Blood, with over 150 proteins, serves as a good example to illustrate the events that occur during interaction of a foreign surface with a multicomponent solution. On the basis of mass transport considerations, proteins present at the highest concentration (Table 3.3) will be first to arrive at the surface. The order of availability according to the simplest form of transport, pure diffusion, is shown in Table 3.4. Because of its high concentration and moderate size (diffusion coefficient), albumin dominates initial interactions with the surface. IgG, with its lower concentration and larger size, has a slower rate of arrival at the

TABLE 3.3. Plasma Proteins With the Highest Concentration

Protein	Concentration (mg/ml)	Molecular Weight	Diffusion Coefficient (10^{-7} cm^2/s)
Albumin	40	66,000	6.1
IgG	15	150,000	4.0
α_1-Antitrypsin	3	54,000	5.2
Fibrinogen	3	340,000	2.0
Low-density lipoprotein (LDL)	3	5,000,000	5.4
α_2-Macroglobulin	3	725,000	2.4
Transferrin	2.6	77,000	5.0
IgA	2.3	162,000	3.4
α_2-Haptoglobins	2	100,000	4.7
High-density lipoprotein (HDL)	2	195,000	4.6
Complement 3	1.6	180,000	4.5

Adapted from Andrade, J.D. and Hlady, V. (1987). Plasma protein adsorption: The big twelve, *Ann. NY Acad. Sci.* 516:158–172.

TABLE 3.4. Rate of Arrival of the Proteins Listed in Table 3.3, Based on Diffusion-Limited Mass Transport Described by *Equation 1*

Protein	C (μM)	D (10^{-7} cm^2/s)	$C\sqrt{D}$
Albumin	606	6.1	1,497
IgG	100	4	200
α_1-Antitrypsin	56	5.2	127
Transferrin	34	5	76
α_2-Haptoglobins	20	4.7	43
IgA	14	3.4	26
High-density lipoprotein (HDL)	10	4.6	22
Complement 3	9	4.5	19
Fibrinogen	9	2	12
α_2-Macroglobulin	4	2.4	6
Low-density lipoprotein (LDL)	1	5.4	1

Adapted from Andrade, J.D. and Hlady, V. (1987). Plasma protein adsorption: The big twelve, *Ann. NY Acad. Sci.* 516:158–172.

surface. In terms of incidence of protein molecules colliding with the surface, seven times less IgG interacts with the biomaterial than does albumin. Almost 1,500 times less low-density lipoprotein than albumin interacts with the surface. Even though IgG has a slower rate of arrival than does albumin, if its molecules have a greater affinity for the surface, IgG molecules can exchange with bound albumin molecules. Similarly, other proteins with slower rates of arrival sequentially arrive at the surface, and depending on their affinity for the biomaterial, they can replace previously adsorbed molecules. Therefore, fibrinogen, for example, can dominate the surface because of greater affinity, even though its rate of arrival is over 100 times less than that of albumin. The actual hierarchy of blood proteins on surfaces, however, is more complicated than this simplified, diffusion-limited example (Table 3.5). Because of their affinity or additional kinetic factors (e.g., convection), molecules other than those with the highest concentration will also bind to the surface. Adsorption of proteins involved in blood clotting, such as fibrinogen and factor XII (discussed in Chapter 4), has great importance for determining tissue-implant interactions.

TABLE 3.5. Exchange Hierarchy of Plasma Proteins on Glass and Metal Oxide Surfaces

Protein	
Albumin	Adsorbs First
IgG	↓
Fibrinogen	↓
Fibronectin	↓
Factor XII	↓
High-molecular-weight kininogen	↓

3.6 SUMMARY

- The interaction of proteins with biomaterials is determined by the properties of both the biomolecules and substrate.
- Protein factors that affect their interaction with biomaterials include size, charge, amphipathicity, and structural stability.
- Surface factors that influence their interaction with proteins include topography, charge, chemical composition, and microstructure.
- The rate of arrival of protein molecules at a biomaterial surface also plays a significant role in determining adsorption.
- The multiple states in which proteins exist on surfaces result from effects of orientation, geometric availability of surface area, and conformational changes.
- All protein-surface bonds must be simultaneously broken for a protein molecule to desorb.
- The longer a protein molecule resides on a surface, the less likely it is to be desorbed or exchanged by other molecules.
- In multicomponent solutions, such as real body fluids, proteins compete for surface binding sites, resulting in a series of collision, adsorption, and exchange processes on the biomaterial surface.

3.7 BIBLIOGRAPHY/SUGGESTED READING

Andrade, J.D. Principles of protein adsorption, *Surface and Interfacial Aspects of Biomedical Polymers*, Vol. 2, J.D. Andrade (ed.), Plenum Press, New York, 1985, pp. 1–80.

Horbett, T.A. and Brash, J.L. Proteins at interfaces: Current issues and future prospects, *Proteins at Interfaces: Physiochemical and Biochemical Studies*, J.L. Brash and T.A. Horbett (eds.), American Chemical Society, Washington, D.C., 1987, pp. 1–33.

3.8 QUIZ QUESTIONS

1. How does the size of a protein influence its ability to bind to a biomaterial?
2. Explain how the presence of disulfide bonds within a protein influences its ability to adsorb to surfaces.
3. How do adjacent protein molecules affect adsorption in a single component solution? In a multicomponent solution?
4. What factors affect the rate of arrival of protein molecules at a surface?
5. Explain the differences between protein desorption and exchange.
6. Why is pure desorption of a protein unlikely?

3.9 STUDY QUESTIONS/DISCOVERY ACTIVITIES

1. How does the roughness of a biomaterial surface affect protein adsorption? Consider roughness at different scales, that is, macro-, micro-, and nano-roughness.

2. Using *Equation 3.2* and the data shown in Table 3.3, plot the number of molecules of albumin and fibrinogen adsorbing on a surface during the first minute of interaction. How would fluid flow, such as caused by micromotion at the tissue-implant interface, alter adsorption?

3. Why is protein removal more difficult the longer it interacts with a surface? Using the information presented in both Chapters 2 and 3, explain this process, beginning with a protein in solution and ending with an adsorbed molecule that is resistant to removal.

4. Describe the Vroman effect. What is the significance of this phenomenon? What factors influence it?

5. Conduct a literature search on the topic of protein adsorption on biomaterials. Is this an active area of research? What is the focus of the citations you found? Select a recent publication, and summarize it in terms of the concepts presented in this chapter.

4

Blood-Biomaterial Interactions and Coagulation

4.1 INTRODUCTION

Implanting a biomaterial or device requires injuring or wounding the patient, and bleeding generally ensues—as a matter of fact, blood is the first "tissue" that the surface of any implanted biomaterial/device will probably contact. It's important to realize that the processes of wound healing and tissue repair are automatically triggered by tissue injury (such as from an implantation procedure). Cell death, disruption/denaturation of the extracellular matrix, and loss of blood vessel integrity are some of the events that can directly trigger wound healing, starting with the first phase: coagulation (Fig. 0.2). This chapter outlines the origins and roles of blood-borne cells and chemicals in coagulation and fibrinolysis.

4.2 THE BLOOD CELL SOURCE: MARROW AND STEM CELLS

Blood is a mixture of *plasma* (a solution of water, salts, and proteins), various kinds of cells, and platelets (biologically active cellular fragments) (Fig. 4.1). All blood cells originate in the bone marrow via a complex process termed *hematopoiesis*. At birth, bone marrow throughout the body is red and produces blood cells, but over time, much of the marrow changes to yellow (adipose) tissue and stops producing blood cells. This change is reversible in times of stress and disease. By adulthood the red, blood-cell producing bone marrow tends to be found in cancellous bone (bone with an open, porous structure) like the ends of the humerae, in the tibiae and femurs, and in portions of the pelvis, ribs, and sternum. Within the red marrow is the *stroma* (from the Greek word for "bed"), a three-dimensional network of spongelike tissue made up primarily of cells and structural fibers (Fig. 4.2). Within the stroma are *sinuses* (empty spaces) and the *sinusoidal capillaries* that provide the blood supply for the marrow cavity. Blood cells are continually being formed in the stroma, squeezing into the sinusoidal

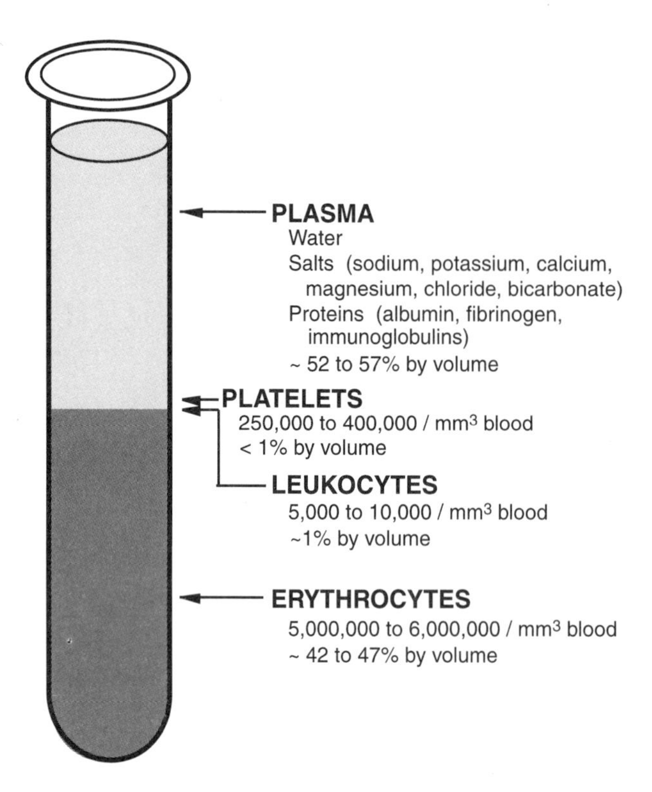

Figure 4.1. Constituents of blood by volume. If this were a real test tube of blood separated into components, the leukocytes would be entirely contained within the small white band underneath the layer of platelets (which would be too thin to be seen easily).

capillaries, and entering the bloodstream to be circulated throughout the rest of the body.

All of the cells that circulate in the bloodstream are derived from one type of cell in the bone marrow: the *pluripotent hematopoietic stem cell. Pluripotent* implies that a cell is capable of replicating and of differentiating into multiple types of cells. The word *hematopoietic* is formed from *hema*, meaning "blood," and *poietic*, meaning "forming." Finally, the general term *stem cell* is used to describe a cell that is capable of continued replication and production of other cell types. Pluripotent hematopoietic stem cells (PHSCs) are *multipotent*, or capable of producing a variety of cells within a given class (for example, blood cells, rather than skin cells), rather than *totipotent*, or capable of producing any other form of cell. PHSCs are relatively rare—it's been estimated that only perhaps 1 in 1,000 bone marrow cells is a PHSC—and are thought to be generally inexhaustible over the lifetime of a person, barring radiation damage or certain diseases. PHSCs and the other hematopoietic cells in the marrow produce enough cells for about an ounce of blood each day, which doesn't sound very impressive until you consider the fact that an ounce of blood contains about 260 billion new cells.

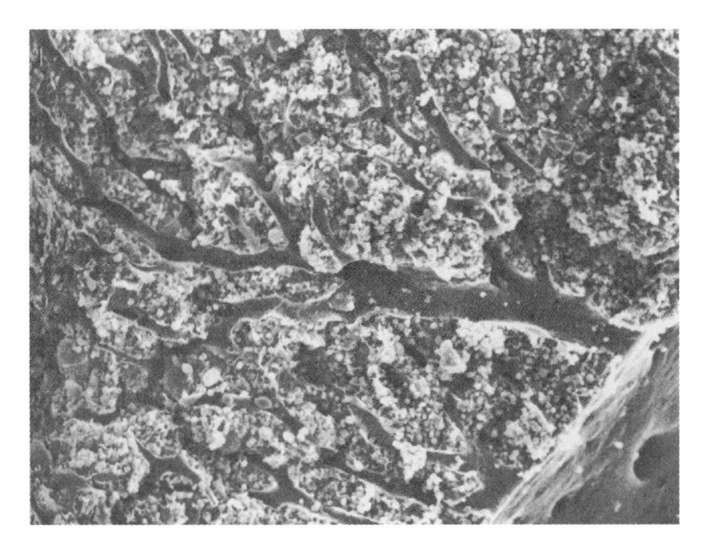

Figure 4.2. Scanning electron micrograph of rat bone marrow. A large vein is seen in the lower right corner, with smaller veins throughout the marrow. The many small, round objects are cells. Original magnification: ×242. Figure reproduced from L. Weiss, "The hematopoietic microenvironment of the bone marrow: an ultrastructural study of the stroma in rats," *Anatomical Record* 186:161–184 (1976), © John Wiley & Sons, Inc., 1976. Reprinted by permission of Wiley-Liss, Inc., a subsidiary of John Wiley & Sons, Inc.

PHSCs can replicate (i.e., produce more PHSCs) and can also differentiate into two new stem cell types: the *lymphoid stem cell* and the *myeloid stem cell.* The lymphoid stem cells migrate from the bone marrow to parts of the lymphatic system (lymph nodes, spleen, thymus), where they eventually produce *lymphocytes*—white blood cells of the lymphatic system. The myeloid stem cells remain in the bone marrow and produce all of the blood-borne cells students are used to hearing about: the oxygen-carrying *red blood cells*, the *platelets*, which play important roles in the process of blood clotting, and the *leukocytes*, which are part of the body's defense against foreign objects (like biomaterials!). The general lineage of these different blood-derived cells is shown in Figure 4.3, with most of the intermediate differentiation steps left out. Many of the missing differentiation steps are supplied in subsequent figures in this book, in discussions of the functions of specific types of blood-derived cells.

4.3 RED BLOOD CELLS

4.3.1 Formation and Function

Red blood cells are the end stage of a long process of cell division and differentiation, shown in Figure 4.4. Although red blood cells play minimal roles in wound healing and in blood-biomaterial interactions, the production and func-

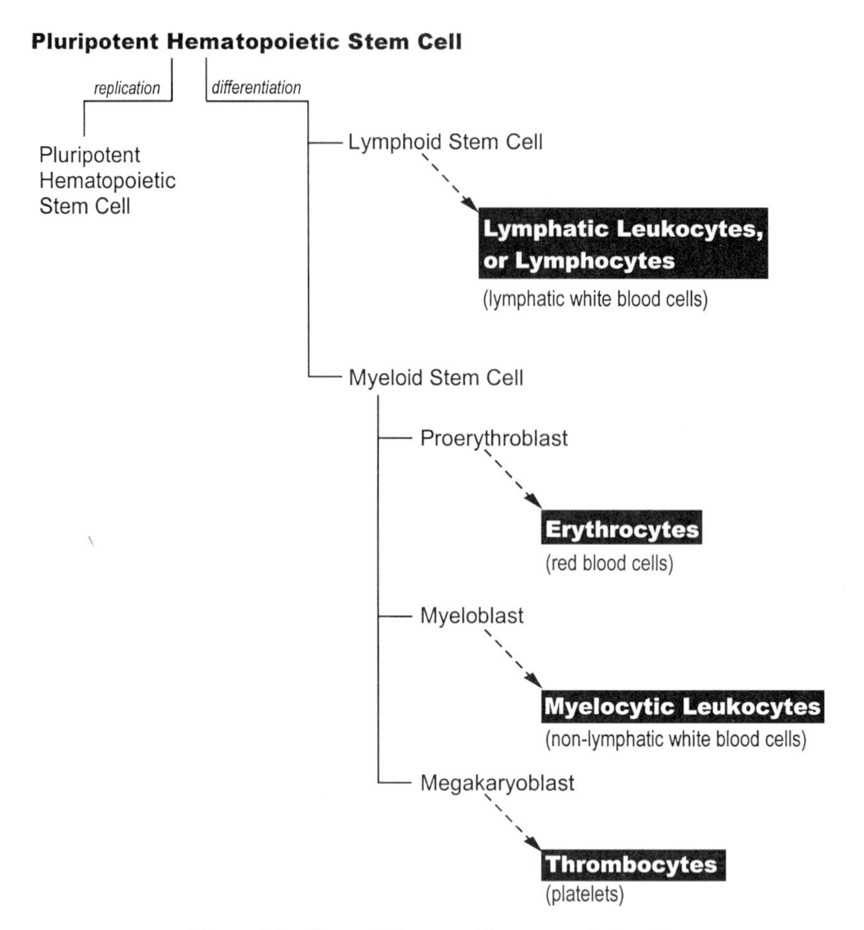

Figure 4.3. General lineage of hematopoietic cells.

tion of these cells can be used to introduce and demonstrate fundamental biological and engineering concepts. The diagram in Figure 4.4 can be used as a sort of "translation guide" for terms commonly used to describe cells. For example, the prefix *erythro* appears almost everywhere in the red blood cell lineage, and for good reason, because it indicates the color red. *Cyte* designates *cell*, whereas *blast* is used to denote a *precursor cell*, or a cell that has the capability to differentiate into another type of cell. The reticulocyte is the only cell in the lineage that doesn't seem to follow the naming pattern; *reticulo* refers to having a *network of filaments*. As the precursor cells near the top of the red blood cell lineage reproduce by dividing, they also differentiate into slightly different types of cells (proerythroblasts become basophil erythroblasts, which in turn become polychromatic erythroblasts, etc.); as this process continues each new cell type has a slightly smaller volume than the precursor cell before, and the nuclear matter condenses to be smaller and smaller. At the orthochromatic erythroblast stage, mitosis no longer occurs; instead, the nucleus is expelled from the cell (often during the process of the cell squeezing into a sinusoidal capillary

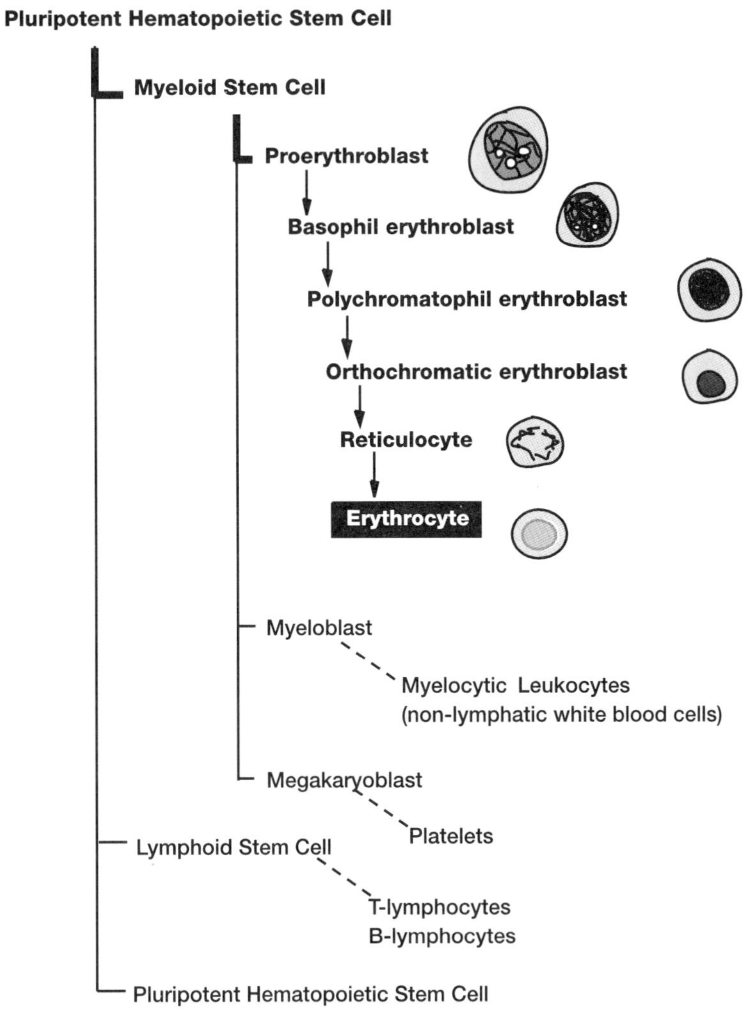

Figure 4.4. Lineage of erythrocytes. The sketches of cells at varying stages illustrate the major structural changes that occur with division and differentiation.

from the stroma), leaving behind a reticulocyte (Fig. 4.5). Reticulocytes contain a small amount of residual nuclear fragments—the filaments for which the cells are named—and can be considered immature erythrocytes. Within a few days, the remaining nuclear fragments as well as any cytoplasmic mitochondria, ribosomes, and other enzymes are denatured and the reticulocyte is then a mature erythrocyte.

Because they have no nucleus or cytoplasmic organelles needed for protein synthesis, erythrocytes do not proliferate and do not synthesize hemoglobin. These cells function purely as a transport mechanism for oxygen and carbon dioxide. Human erythrocytes can survive in the circulation for about 120 days before they wear out and are removed—usually by specialized cells in the spleen

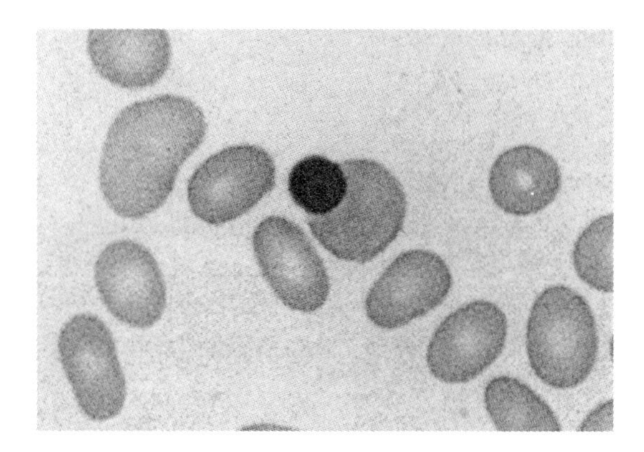

Figure 4.5. Production of a reticulocyte. The erythroblast is in the process of extruding its nucleus (dark circle in center). The nucleus will degrade and/or be destroyed by scavenger cells, and the remaining cytoplasm will be a reticulocyte. The other cells visible are erythrocytes. Figure reproduced from F.G.J. Hayhoe and R.J. Flemans, *Color Atlas of Hematological Cytology*, page 33, © Mosby-Year Book, Inc., 1992. Reprinted by permission of W.B. Saunders Company/Mosby/Churchill Livingstone.

and bone marrow that recognize and consume dead cells or foreign particles (these specialized scavenger cells are *macrophages*; they are discussed in Chapter 5). Therefore, to meet the oxygen demands of the body, the process of red blood cell generation is continually occurring and can be upregulated in response to increased oxygen demand.

A number of factors that result in poor tissue oxygenation (such as low blood volume, anemia, pulmonary disease, or low hemoglobin levels) stimulate the production of red blood cells by increasing the levels of *erythropoietin* (Fig. 4.6).

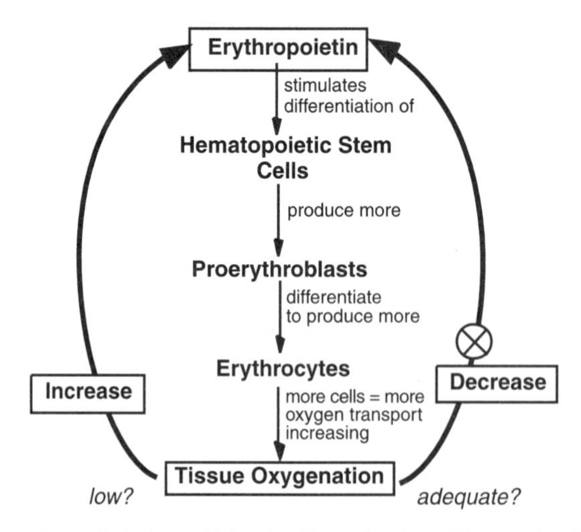

Figure 4.6. Role of erythropoietin in red blood cell production. The production of erythropoietin, and therefore of erythrocytes, depends on tissue oxygenation levels.

Erythropoietin is mainly formed in the kidneys, and the production of erythropoietin can be upregulated within minutes to hours of placing a person in a low-oxygen environment. The erythropoietin stimulates the production of proerythroblasts from hematopoietic stem cells in the bone marrow, and new red blood cells reach the circulation about 5 days after the initial low-oxygen stimulus. Erythropoietin will continue to be produced, causing increased red blood cell production, until the tissue oxygenation level returns to normal.

4.3.2 Deformation and Blood Flow

Under normal conditions, erythrocytes are extremely deformable—they have no nucleus and are essentially a drop of liquid, containing hemoglobin and a few enzymes, wrapped in a lipid bilayer (the cell membrane). Typically, erythrocytes are biconcave and about 7 μm in diameter (Fig. 4.7). The biconcave shape maximizes the ratio of surface area to volume and, if the cell membrane provides negligible resistance to bending, implies equilibrium of the pressures inside and outside the red blood cell membrane. Some of the capillaries in the human body have diameters of 5–7 μm, so erythrocytes deform dramatically to fit through these small vessels (Fig. 4.8). The deformation of erythrocytes plays a large role in determining fluid properties of blood. Normal blood plasma behaves like a Newtonian fluid, for which shear applied to the fluid (τ) is related to the resulting velocity gradient within the fluid (du/dy) by the viscosity coefficient μ:

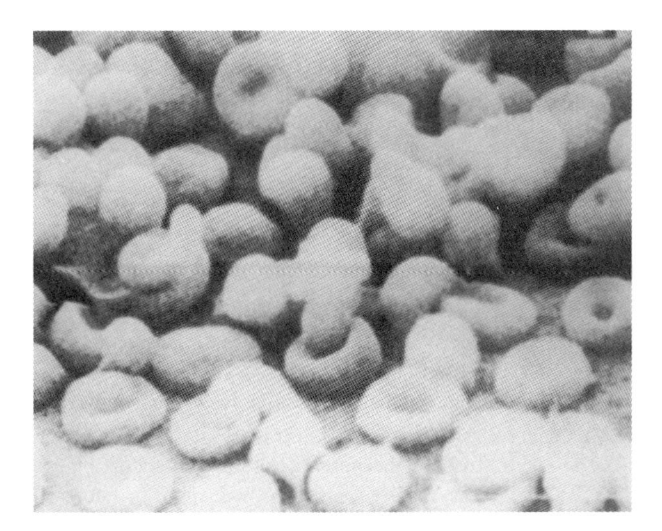

Figure 4.7. Micrograph of erythrocytes. These cells are adhering to a surface of titanium oxide, and most of the cells exhibit a typical, biconcave morphology. Figure reproduced from F. Zhang, Z. Zheng, Y. Chen, X. Liu, A. Chen, and Z. Jiang, "In vivo investigation of blood compatibility of titanium oxide films," *Journal of Biomedical Materials Research* 42:128–133 (1998), © John Wiley & Sons, Inc., 1998. Reprinted by permission of Wiley-Liss, Inc., a subsidiary of John Wiley & Sons, Inc.

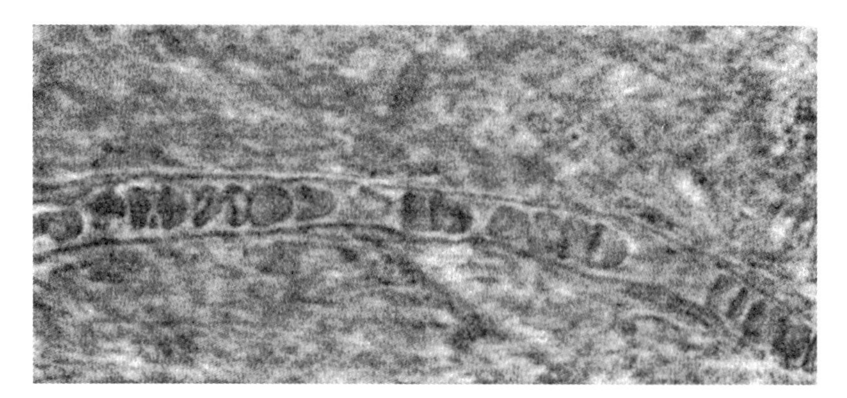

Figure 4.8. Erythrocytes deformed by flow. These cells are flowing from the left to the right, through a capillary in a rabbit ear. ×850 original magnification. Figure reproduced from W.J. Cliff, *Blood Vessels*, page 15, © Cambridge University Press, 1976. Reprinted with the permission of Cambridge University Press and of the author.

$$\tau = \mu \frac{du}{dy} \quad \text{(see Fig. 4.9)}$$

The viscosity of a Newtonian fluid like blood plasma is a thermodynamic property and varies with temperature and pressure. Whole blood (which includes

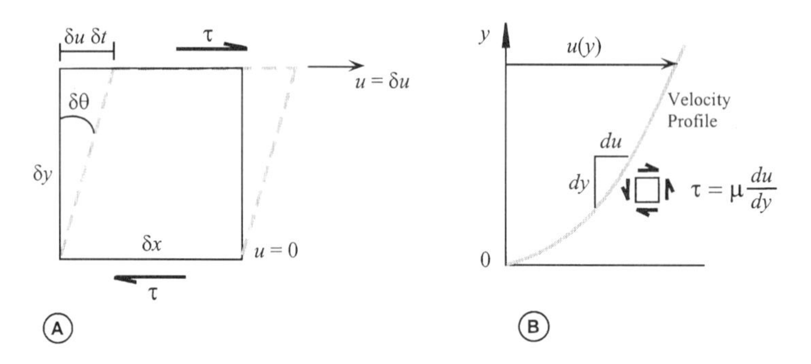

Figure 4.9. Shear stress causing shear deformation in a fluid. (**A**) A fluid element (δx by δy) is sheared in one plane by a shear stress τ, thereby being strained at a rate of $\delta\theta/\delta t$. The top of the element will move at a speed u $= \delta u$ greater than the bottom of the element (where $u = 0$), resulting in a shear deformation at the top of $\delta u \times \delta t$. (**B**) Shear distribution in a fluid boundary layer near a channel wall. Instead of focusing on the strain angle θ, fluid mechanics focuses on the velocity profile $u(y)$. The shear stress in the boundary layer is proportional to the slope of the velocity profile, du/dy. The shear stress is greatest at the wall (note the steep slope of the velocity profile near $y = 0$), where the velocity of the fluid is zero—that is, where the "no-slip condition" is applied. The applied shear τ is related to the slope of the velocity profile via the viscosity coefficient, μ. For Newtonian fluids, the viscosity coefficient does not depend on the rate at which the shearing strain is applied, but only on the temperature and pressure of the fluid. Because generally the viscosity of a fluid changes only slightly as the pressure changes, many engineering analyses neglect the effects of pressure and only consider temperature. Figure adapted and redrawn from F.M. White, *Fluid Mechanics* (third edition), McGraw-Hill, Inc., New York, page 20 (1994).

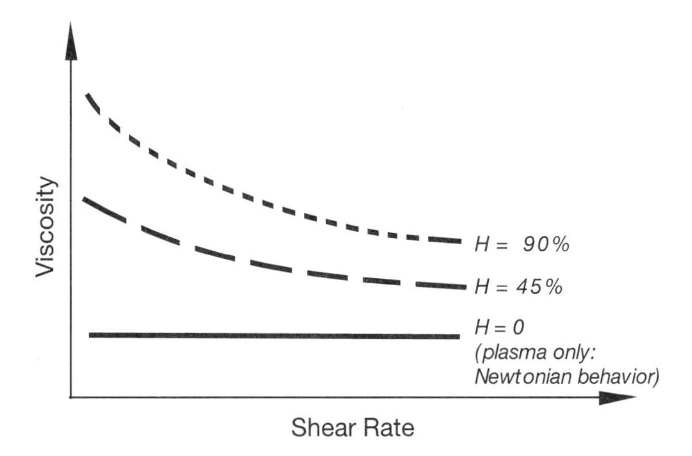

Figure 4.10. Viscosity of blood as a function of hematocrit (H). Plasma alone (with no erythrocytes) behaves like a Newtonian fluid in that the viscosity does not depend on shear rate. With increasing hematocrit, the viscosity-shear rate dependence increases. Simplified and redrawn from Fung, Y.C., *Biomechanics: Mechanical Properties of Living Tissues*, second edition, Springer-Verlag, New York, NY (1993), p. 67.

plasma and the cellular components of blood) is a non-Newtonian fluid, because the viscosity of whole blood varies with the applied shear rate as well as with the *hematocrit*, or percentage of cells within the blood (Fig. 4.10). The non-Newtonian characteristics of blood flow are due to the flow behavior of red blood cells. At very low flows (and correspondingly low shear rates), human red blood cells stack into aggregates called *rouleaux* (Fig. 4.11); as the shear rate approaches zero, the rouleaux aggregate into masses, and blood essentially behaves as a solid. When the shear rate increases, the relatively weak protein-

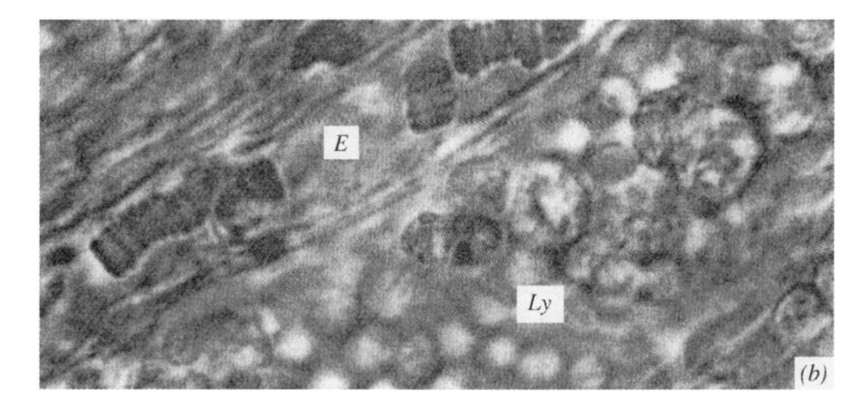

Figure 4.11. Rouleaux formation in a slow-flowing blood vessel. The stacks of erythrocytes are visible in the vessel marked "*E*". "*Ly*" denotes a lymphatic vessel, containing macrophages and leukocytes, directly next to the blood vessel. Figure reproduced from W.J. Cliff, *Blood Vessels*, page 16, © Cambridge University Press, 1976. Reprinted with the permission of Cambridge University Press and of the author.

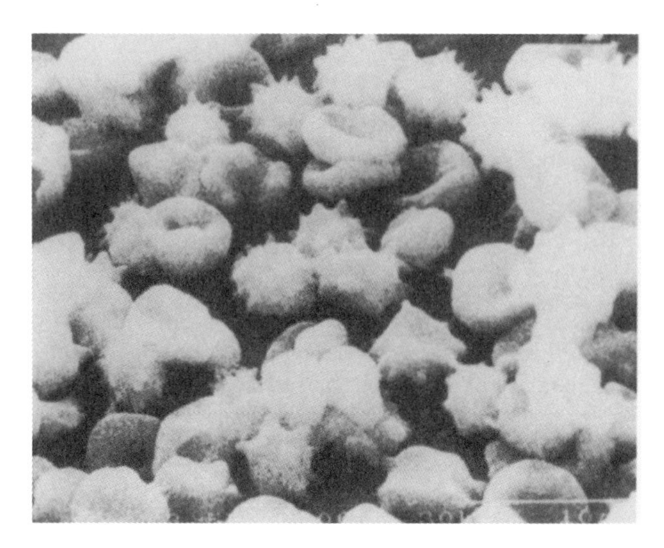

Figure 4.12. Distorted erythrocytes. These cells are adherent to low-temperature isotropic pyrolytic carbon, a biomaterial that has been used to fabricate artificial heart valves. Figure reproduced from F. Zhang, Z. Zheng, Y. Chen, X. Liu, A. Chen, and Z. Jiang, "In vivo investigation of blood compatibility of titanium oxide films," *Journal of Biomedical Materials Research* 42:128–133 (1998), © John Wiley & Sons, Inc., 1998. Reprinted by permission of Wiley-Liss, Inc., a subsidiary of John Wiley & Sons, Inc.

membrane interactions that hold rouleaux together are broken apart, and the viscosity of the blood decreases. At higher shear rates, the erythrocytes become elongated and align with the fluid streamlines, further reducing the apparent viscosity of blood. The biconcave shape of the erythrocyte allows the cells to deform quite a bit without increasing the total surface area, and thus without increasing the surface energy.

Changes in erythrocyte shape (Fig. 4.12) can be caused by changes in pH and *osmolarity*, as well as by various pathological conditions. Osmolarity is illustrated in Figure 4.13 as a tendency for water to move across a semipermeable membrane to establish a solute concentration equilibrium; in actuality, the fluid movement occurs to equilibrate the chemical potential. Osmolarity is similar to molarity in that both concepts are based on the number of particles of a solute in a solution rather than on the mass of the solute in a solution. However, osmolarity accounts for solutes that will dissociate into multiple particles in a solvent. For example, one mole of NaCl placed in one liter of water creates a 1 molar solution. One mole of NaCl placed in one liter of water creates a 2 osmolar solution, however, because the NaCl will dissociate into two ions: Na^+ and Cl^-. Charge isn't accounted for in the concept of osmolarity, so osmotically, a Cl^- ion is equivalent to a Ca^{2+} ion. If a living cell is placed into a solution with a higher concentration of solutes or higher osmolarity than physiological levels, that *hypertonic* solution causes water to pass outward through the cell membrane. This loss of water can cause the cell to shrink (*crenate*). If a

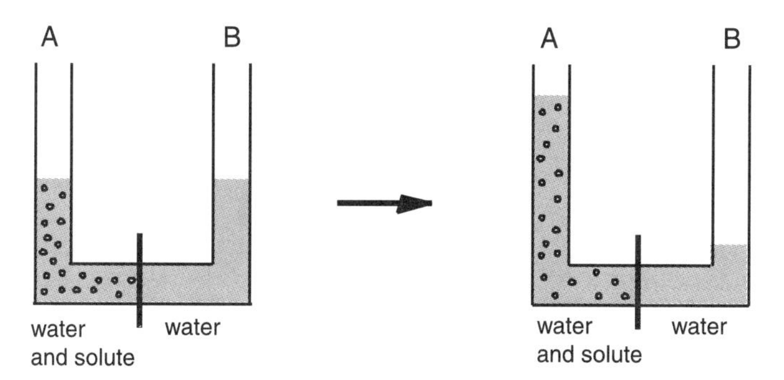

Figure 4.13. Schematic illustration of the concept of osmotic pressure. The semipermeable membrane allows water to pass freely but retains the solute on side A. Over time, water moves from side B to side A of the membrane, to equilibrate the chemical potential across the membrane. However, this flow of water increases the height, and thus the weight (solution density × gravitational acceleration × column height), of the column of water/solute on side A. When the system has come to equilibrium such that the (osmotic) tendency of water to flow from side B to side A has been balanced by the (gravitational) tendency of water to flow from side A to side B, the difference in height between columns A and B is the effective osmotic pressure.

living cell is placed into a solution with a lower concentration of solutes or lower osmolarity than physiological levels, that *hypotonic* solution causes an influx of water into the cell, and the cell expands. If the cell swells too much or too rapidly, the membrane will burst (*lyse*) and the cell will die. When working in a laboratory, care is taken to expose living cells and tissue to *isotonic* saline solutions (for example, phosphate-buffered saline, a solution so ubiquitous in a cell laboratory that often it is referred to simply as "PBS") instead of water. Water is hypotonic compared to the interior of a cell, so immersing living cells in plain water is a good way to disrupt the cell membrane. In the case of erythrocytes, osmotic swelling bulges the membrane outward toward a spherical shape, until the cell bursts.

The normal, flexible membrane shape of erythrocytes is crucial to their function: If they were not so deformable, they would be unable to fit through small capillaries, blocking blood flow and causing oxygen shortage (*hypoxia*) in tissues. Unfortunately, this is exactly what happens in *sickle-cell disease*. Sickle-cell disease is an inherited disorder affecting hemoglobin. The hemoglobin molecule is comprised of four hemoglobin chains, each of which consists of an iron-containing and oxygen-binding heme group and a long *polypeptide* (chain of amino acids). Hemoglobin chains are classified as alpha, beta, gamma, or delta chains according to the composition of the polypeptide; the most common form of hemoglobin in adults consists of two alpha and two beta chains. Sickle-cell disease is caused by a single mutation in the DNA of the gene that codes for the polypeptide of the beta chain of hemoglobin. Specifically, the triplet GAA, which codes for glutamic acid, is replaced by the triplet GUA, which codes for valine. As a result, as hemoglobin is synthesized by cells through the erythrocyte

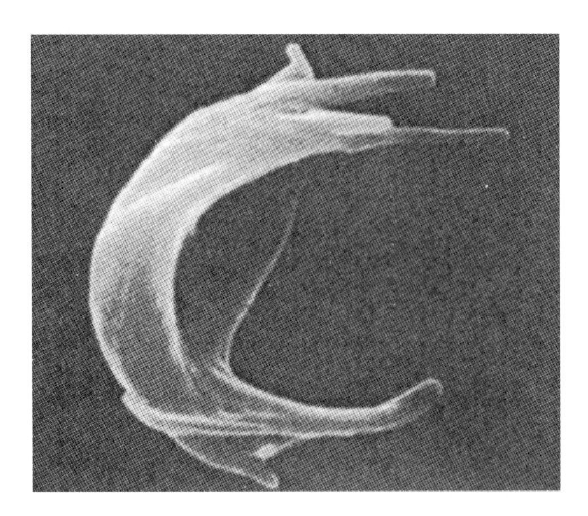

Figure 4.14. Distorted erythrocyte due to sickle-cell disease. The rigid hemoglobin inside the erythrocyte prevents easy deformation of the cell, blocking blood flow through smaller vessels, and potentially puncturing the cell membrane. Figure reproduced from L.C. Junquiera, J. Carneiro, and R.O. Kelley, *Basic Histology*, eighth edition, page 221, © Appleton & Lange, 1995, with the permission of The McGraw-Hill Companies.

lineage (from proerythroblasts down to reticulocytes), beta chains are synthesized with one amino acid misplaced: a valine appears where a glutamic acid should be. This substitution causes the hemoglobin to polymerize under low-oxygenation conditions, forming elongated crystals within the red blood cells that can be up to 15 μm in length (Fig. 4.14). These crystals are inflexible, often causing the "sickled" erythrocytes to block capillaries and small blood vessels. The sharp, pointed crystals also have the potential to rupture the erythrocyte membrane. If enough red blood cells rupture from the polymerized hemoglobin, the overall concentration of active hemoglobin in the blood may be significantly decreased: this is known as sickle-cell disease. Sickle-cell disease is an example of how very small changes in chemistry, and in the subsequent structure and function of a protein, can lead to dramatic physiological effects.

4.4 PLATELETS

4.4.1 Formation and Function

Platelets originate in the bone marrow, via the fragmentation of larger cells (Fig. 4.15) derived from megakaryoblasts. Megakaryoblasts (*mega* meaning large and *karyo* implying the nucleus), which can range from 15 to 50 μm in diameter, differentiate to multiply the amount of DNA in the nucleus by about thirty times before becoming a megakaryocyte. Megakaryocytes are big cells, from 35 to 150 μm in diameter. The cytoplasm of megakaryocytes contains a large quantity of well-developed rough endoplasmic reticulum, an extensive

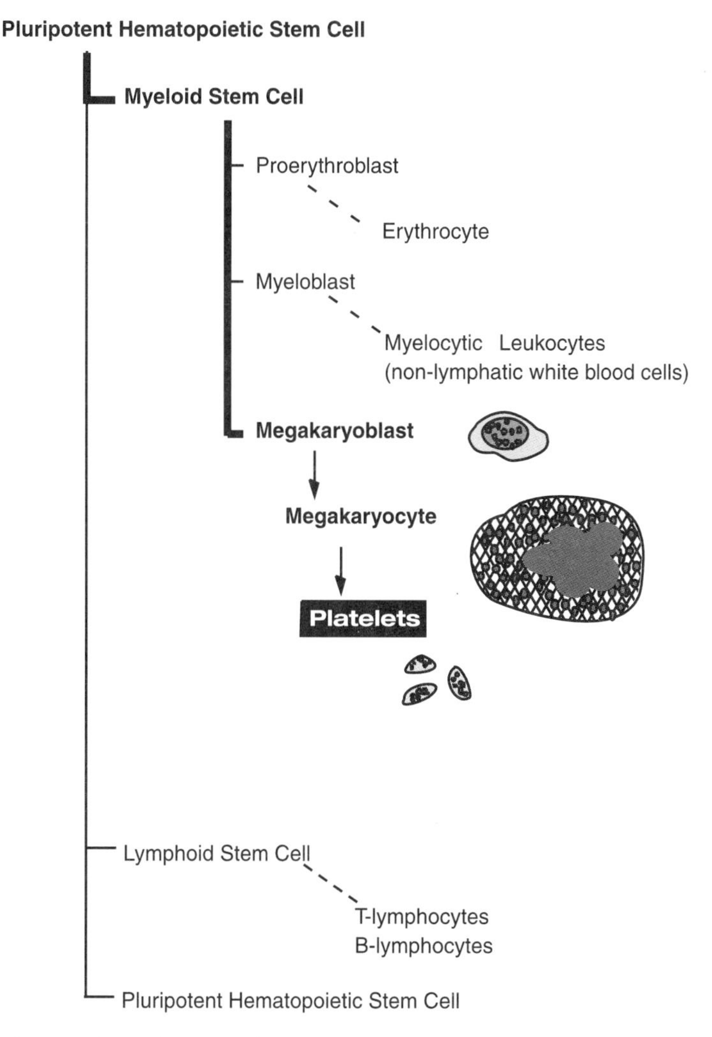

Figure 4.15. Lineage of platelets. Platelets are cytoplasmic fragments formed from the disruption of megakaryocytes.

Golgi complex, and numerous mitochondria. As the cell matures, the cell membrane invaginates and develops a network throughout the cytoplasm. These membranes are called *demarcation membranes*. The megakaryocyte fractures along the demarcation membranes because of the stresses experienced when squeezing into the marrow vasculature or, once within the vasculature, when pushed into small-diameter capillaries. The small fragments of megakaryocyte cytoplasm contained within the demarcation membranes, only 2–4 μm in diameter, are platelets (Fig. 4.16).

Platelets have no nucleus and thus cannot proliferate; after a half-life of about 8–10 days, inactive platelets are cleaned out of the spleen by macrophage scavenger cells. Figure 4.17 is a micrograph of human platelets, showing round

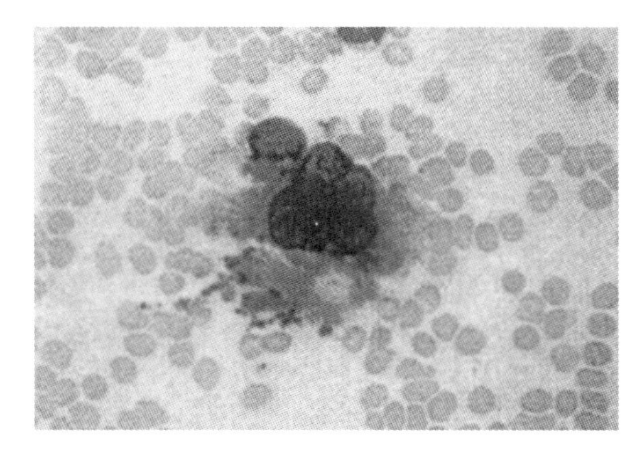

Figure 4.16. Platelet formation. The megakaryocyte in the center of this figure is in the process of breaking into many platelets (tiny dark spots in the figure). The other cells visible are erythrocytes. Figure reproduced from F.G.J. Hayhoe and R.J. Flemans, *Color Atlas of Hematological Cytology*, page 150, © Mosby-Year Book, Inc., 1992, with the permission of W.B. Saunders Company/Mosby/Churchill Livingstone.

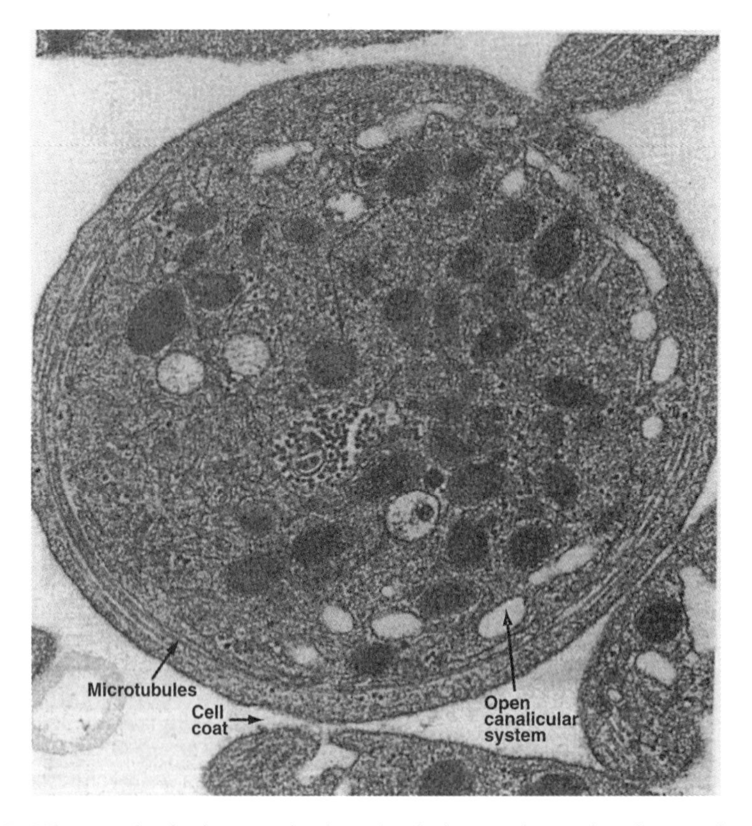

Figure 4.17. Micrograph of a human platelet. The dark, round cytoplasmic granules are visible. Figure reproduced from L.C. Junquiera, J. Carneiro, and R.O. Kelley, *Basic Histology*, eighth edition, page 232, © Appleton & Lange, 1995, with the permission of W.B. Saunders Company/Mosby/Churchill Livingstone.

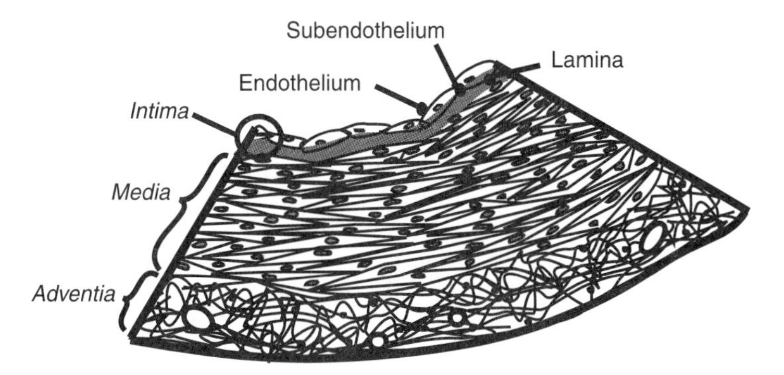

Figure 4.18. Schematic of a cross section through an artery wall. A monolayer of endothelial cells covers the blood-contacting inner surface of the artery; the subendothelium is a layer of loose connective tissue. The lamina can be considered an elastin-containing porous membrane that separates the endothelium from the media while allowing nutrient diffusion into the blood vessel wall. The endothelium, subendothelium, and lamina are referred to together as the intima of the blood vessel. The media is primarily concentric layers of smooth muscle cells, with some elastic fibers, and the adventitia consists of oriented collagen and elastin fibers. Small blood vessels (open circles in the schematic) can pass through the adventitia.

granules as a prominent feature. The granules contain a variety of chemicals that are important to the process of coagulation. Platelets also contain contractile proteins such as actin, myosin, and thrombosthenin. The residual megakaryocyte endoplasmic reticulum, Golgi apparatus, and mitochondria provide a short-term source of intracellular enzymes, ATP, and calcium for the platelet to use. These residual features allow the platelet to be a functionally active and responsive unit, in contrast to the erythrocyte, which is primarily a passive transport mechanism—even though both the platelet and the erythrocyte are essentially cytoplasmic by-products of the destruction of other precursor cells.

One of the platelet's main roles in the vasculature is essentially to "plug holes" in blood vessels by adhering to and covering sites of injury, preventing blood loss from the vessel. Platelets do not adhere to normal, healthy endothelial cells lining the vasculature (Fig. 4.18) but will adhere to exposed connective tissues in injured blood vessels—as well as to the surfaces of many man-made biomaterials. Because adhesion is such a crucial function for these cellular fragments, the platelet membrane possesses glycoproteins and specialized adhesion receptors (for example, the glycoprotein IIb/IIIa receptor pair). This receptor can interact with a specific *peptide*, or sequence of amino acids: arginine–glycine–aspartic acid. This amino acid sequence is part of the amino acid sequence of collagen, a protein in the basement membrane protein that normally would be covered by a layer of healthy endothelial cells. When the endothelium is injured or compromised, and collagen is exposed to the bloodstream, platelets adhere to the collagen via their membrane receptors. In addition, some of the proteins in the blood serum bind to basement membrane components with high affinities: von Willebrand factor, for example, binds to collagen with a high affinity; platelets are then able bind to von Willebrand factor via at least two types of membrane receptors. Platelets can also adhere to other circulating serum proteins, after they adsorb to the surface of a biomaterial and change conformation. Once platelets adhere to a surface, a cascade of events follows, leading to the formation of a "platelet plug" and initiating the process of *coagulation*, or blood clotting.

4.4.2 Platelet Aggregation and the Process of Coagulation

Small disruptions (injuries) of blood vessels happen all the time, through the little and often unnoticed bumps, strains, and injuries that are experienced by human beings going through an ordinary day. When injured endothelial cells and subendothelial structures such as collagen fibers are exposed to blood, platelets adhere to these indicators of damage via cell membrane receptors (Fig. 4.19). Once a platelet adheres to a surface, the contractile proteins within the platelets tighten, and the platelet flattens and forms "legs" or *pseudopodia*, covering more surface area than would be possible if the platelet hadn't changed shape (Figs. 4.20 and 4.21). At the same time (within seconds), the contraction causes the platelets to *degranulate*, or release the contents of the intracellular granules shown in Figure 4.17. Degranulation releases a variety of chemicals, including ADP and thromboxane A_2. These two chemicals are potent activators of platelets, stimulating increased expression of cell membrane adhesion receptors such as glycoprotein IIb/IIIa, and thus nearby platelets are recruited to attach to the adherent platelets and the injury site. These newly adherent platelets degranulate in turn, continuing the cascade of platelet *aggregation* at the site of injury and eventually covering the injured area with a *platelet plug*. Additional contraction of the actin and myosin in the amassed platelets helps to compact and stabilize this plug, drawing the edges of the injury together. Moreover, some of the chemicals secreted by the aggregating platelets are involved in the process of coagulation. An end product of coagulation is the production of sticky threads of a protein called *fibrin*. Fibrin threads attach to and help consolidate the platelet plug as well as trap nearby erythrocytes; at this point the platelet plug has now become a blood clot (Figs. 4.22 and 4.23).

There are two important general mechanisms by which chemicals released from aggregating platelets assist the formation of a blood clot. First, some chemicals produced by the *arachidonic acid cascade* (Fig. 4.24) within the platelet influence subsequent platelet aggregation and formation of the platelet plug. The arachidonic acid cascade is stimulated by the presence of chemicals released from other platelets or which are part of the coagulation process. Some of the products of the arachidonic acid cascade easily penetrate the membranes of other platelets and promote more platelet aggregation, whereas some arachidonic acid cascade products inhibit platelet aggregation. Cyclooxygenase is a key "control point" of the arachidonic acid cascade; the actions of cyclo-

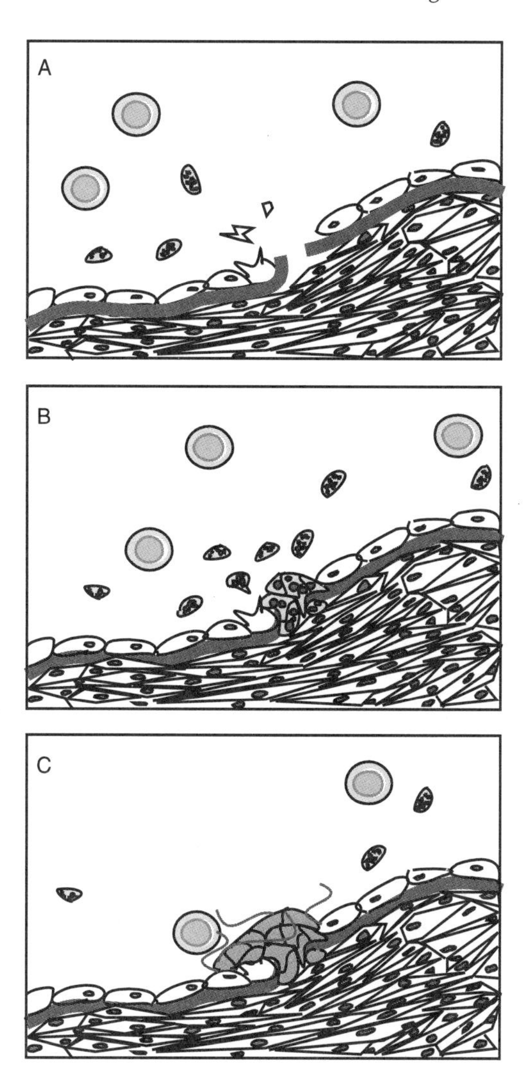

Figure 4.19. Schematic of platelet plug formation. In **A**, the endothelium and the lamina have been compromised, exposing an injured endothelial cell, subendothelium, lamina, and underlying smooth muscle cells to the blood (in which platelets and erythrocytes are shown). In **B**, the platelets have started to adhere to the exposed collagen fibers as well as to the injured endothelial cell and exposed smooth muscle cell. The platelets are now irregularly shaped, and the granules of the adhering platelets have become more prominent in preparation for degranulation and continued platelet aggregation at the injury site. In **C**, the platelets have covered the injury site. The granules are no longer visible because degranulation occurred during the aggregation process. Fibrin threads, formed at the site of the platelet plug as a result of concurrent chemical reactions, are binding and consolidating the plug and have also trapped a nearby erythrocyte.

oxygenase are inhibited by aspirin and other nonsteroidal anti-inflammatory agents. This is one reason why high concentrations of aspirin can impair coagulation for short periods of time, and why aspirin may be given to patients who are at risk of having a heart attack (an occlusion of a blood vessel to the heart)

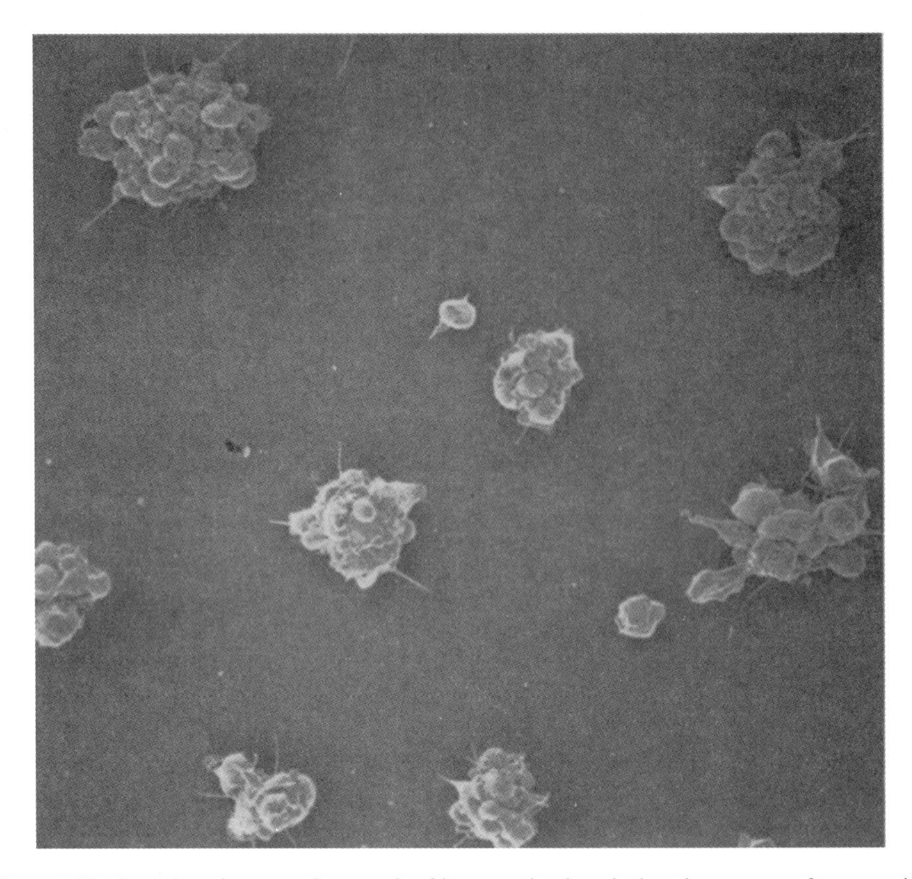

Figure 4.20. Scanning electron micrograph of human platelets during the process of aggregation. The platelets shown are adhering in clusters to Formvar, a commercially available polyvinyl formal resin surface, sometimes used as a cell culture material for imaging studies. Figure reproduced from S.L. Goodman, "Sheep, pig, and human platelet-material interactions with model cardiovascular biomaterials," *Journal of Biomedical Materials Research* 45:240–250 (1999), © John Wiley & Sons, Inc., 1999. Reprinted by permission of Wiley-Liss, Inc., a subsidiary of John Wiley & Sons, Inc.

to prevent a blood clot from forming. The second mechanism by which aggregating platelets assist coagulation is shown by the fact that some platelet-derived factors are used in the production of fibrin threads, as shown in Figure 4.25. Factor V can be released by platelets and is activated (designated with a lower-case "a" in Fig. 4.25) by contact with an enzyme called *thrombin*. Factor Va and factor Xa (which is a chemical product from other mechanisms of coagulation) combine to create a compound called prothrombin activator, which acts to split the plasma protein *prothrombin* into fragments, generating more thrombin. The enzyme thrombin acts to cleave peptide fragments away from another plasma protein, *fibrinogen*, producing *fibrin monomers*. The fibrin monomers polymerize with the help of a *stabilizing factor* (also known as

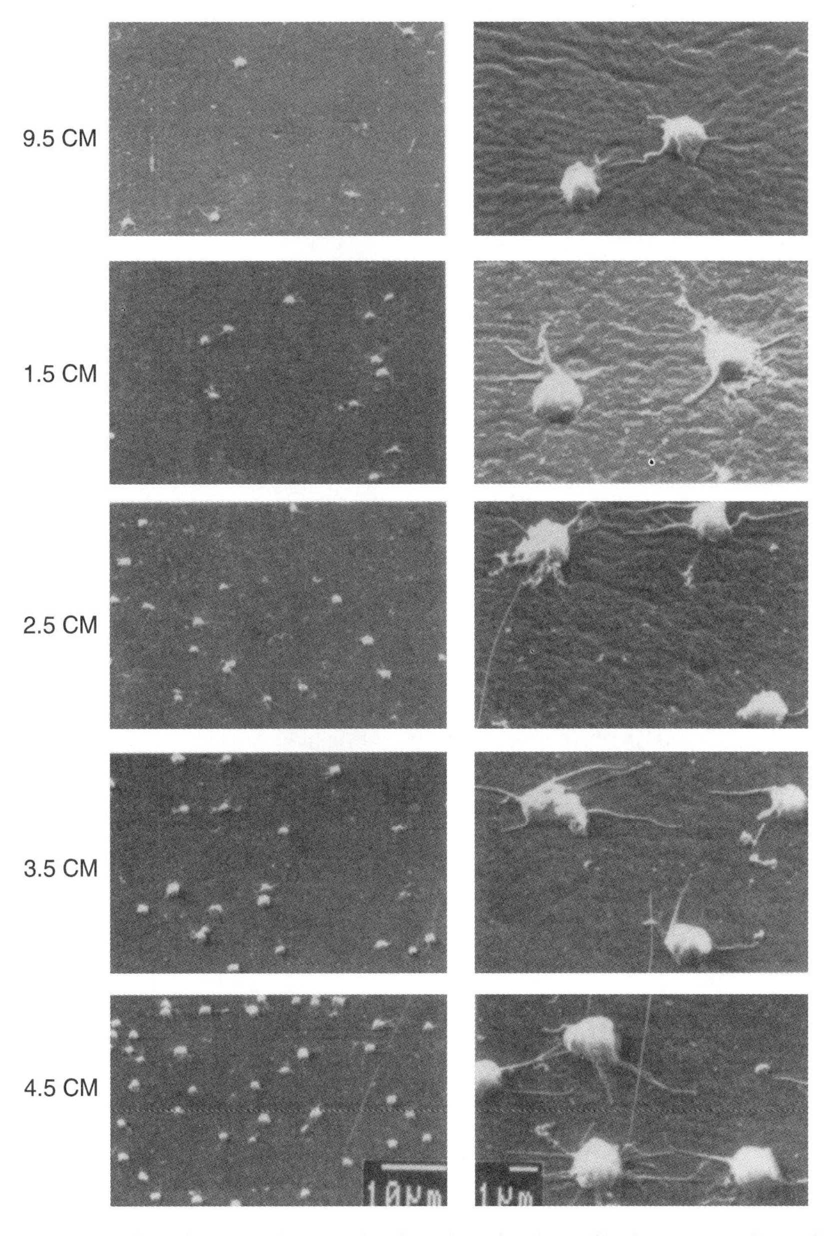

Figure 4.21. Scanning electron micrograph of canine platelets adhering to a surface of varying hydrophilicity. This figure shows typical numbers (left column of pictures) and morphologies (right column of pictures) of platelets adhering at different spatial positions (numbers at left) along a piece of polyethylene in the absence of blood serum proteins. The polyethylene surface had been treated to produce an increasing amount of oxygen-containing functional groups—and thus an increasing hydrophilicity—in a spatial gradient from the top to the bottom of the figure. The more hydrophilic the polyethylene surface, the more the platelets adhered and extended pseudopodia. Figure reproduced from J. Lee and H. Lee, "Platelet adhesion onto wettability gradient surfaces in the absence and presence of plasma proteins," *Journal of Biomedical Materials Research* 41:304–311 (1998), © John Wiley & Sons, Inc., 1998. Reprinted by permission of Wiley-Liss, Inc., a subsidiary of John Wiley & Sons, Inc.

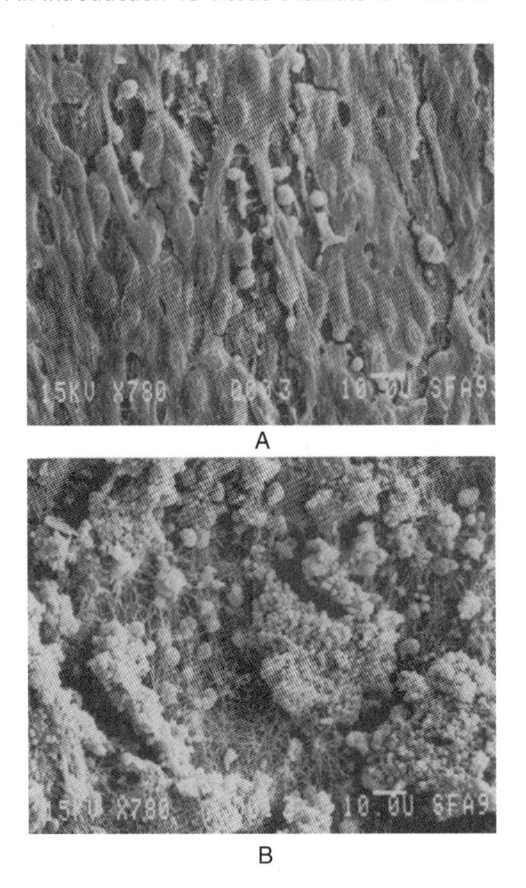

Figure 4.22. Scanning electron micrographs of porcine endothelium. **A** shows a section of undamaged artery; **B** shows a blood clot formed on a section of damaged artery. The dense network of small thread-like strands is fibrin. A large number of tiny, aggregated platelets can be seen in clusters along the biomaterial surface. Larger erythrocytes, having been trapped in the clot as it formed, are also evident. Figure reproduced from Y. Mehri, M. King, and R. Guidoin, "Acute thrombogenicity of intact and injured natural blood conduits versus synthetic conduits: neutrophil, platelet, and fibrin(ogen) adsorption under various shear-rate conditions," *Journal of Biomedical Materials Research* 34:477–485 (1997), © John Wiley & Sons, Inc., 1997. Reprinted by permission of Wiley-Liss, Inc., a subsidiary of John Wiley & Sons, Inc.

factor XIII) which is released by platelets and activated by thrombin to form polymerized fibrin threads.

Two chemicals introduced above as key participants in the production of fibrin threads also play important roles in the process of platelet aggregation. Fibrinogen is necessary for the adhesion of platelets to other platelets and subsequent aggregation. Thrombin also increases platelet-platelet adhesion and, additionally, stimulates platelet activation and degranulation—resulting in the production of even more thrombin. Therefore, the chemical reactions schematically diagrammed in Figure 4.25 are tightly linked to the formation of the platelet

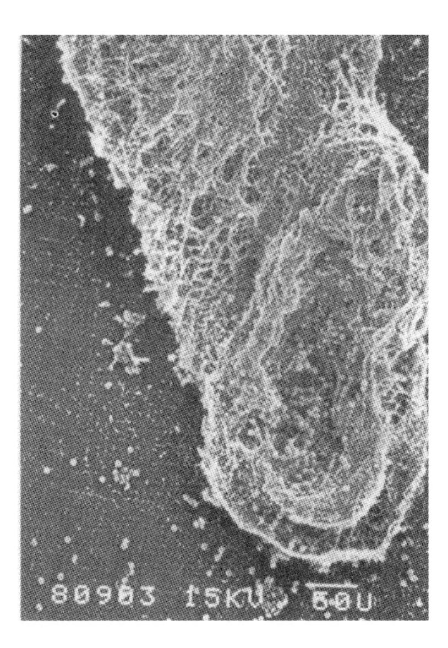

Figure 4.23. Scanning electron micrograph of a clot formed on the surface of a catheter. Blood compatibility of catheter materials is important, even though many procedures involving catheterization of a blood vessel are relatively short. Fluid shear from the flowing blood and mechanical bumping/scraping of a catheter along the lumen of a vessel could dislodge clots formed on a catheter surface, creating potentially dangerous emboli. The catheter shown in this figure was made from polyurethane surface-treated with ozone and acrylic acid. Figure reproduced from H. Inoue, K. Fujimoto, Y. Uyama, Y. Ikada, "Ex vivo and in vivo evaluation of the blood compatibility of surface-modified polyurethane catheters," *Journal of Biomedical Materials Research* 35:255–264 (1997), © John Wiley & Sons, Inc., 1997. Reprinted by permission of Wiley-Liss, Inc., a subsidiary of John Wiley & Sons, Inc.

plug, with a few key constituents (such as thrombin, fibrinogen, calcium, platelet membrane receptors) being absolutely crucial to the formation of a blood clot.

4.5 THE COAGULATION CASCADES

4.5.1 Mechanisms

Have you ever wondered why a small cut from a sharp knife or razor can bleed copiously, whereas, in contrast, a larger wound caused by blunt trauma (shutting one's finger in a door, for example) can stop bleeding quickly? The answer lies in understanding how trauma to tissues and blood vessels initiates and controls the process of blood coagulation, or understanding the concepts of *vascular spasm* and the *extrinsic coagulation pathway*.

Vascular spasm is a constriction of blood vessels, which stops or dramatically slows blood flow. This contraction of the smooth muscle in the injured wall is

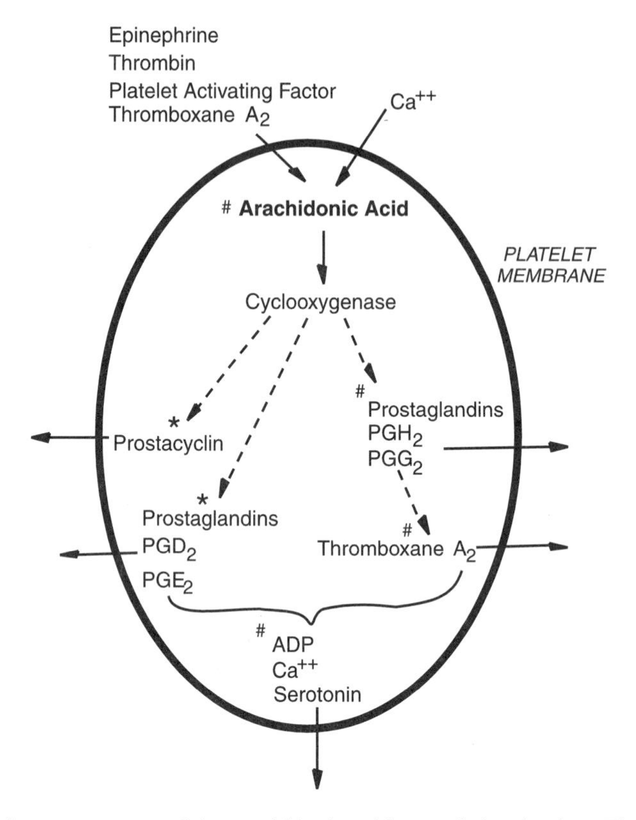

Figure 4.24. Major components of the arachidonic acid cascade in platelets. The factors listed at the top of the diagram pass through the platelet membrane to intracellularly activate the arachidonic acid cascade. Some of the main products of the cascade can easily pass out through the membrane of the platelet and into other platelets. Some of these products (marked with #) promote further platelet aggregation, whereas other products (marked with *) inhibit further platelet aggregation. ADP, Ca^{2+}, and serotonin are by-products of the arachidonic acid cascade and can also penetrate platelet membranes to promote further aggregation.

stimulated by nervous reflexes, probably associated with pain impulses stemming from the injury, and augmented by short-acting metabolites such as endothelin. In smaller blood vessels, chemicals released by damaged or activated platelets—such as thromboxane A_2—act as local vasoconstrictors as well. More vessel trauma means more vascular spasm, reducing blood flow to the site of the injury and slowing bleeding.

Trauma to tissues and blood vessels also initiates blood clotting, by setting one of the coagulation pathways into motion. Coagulation is a result of cascading chemical reactions of plasma proteins (*clotting factors*). These normally inactive factors circulate through the blood and are activated through enzymatic cleavage and/or surface contact (with, for example, the membranes of activated platelets or the surface of a biomaterial). The *extrinsic pathway* is a coagulation cascade that begins with trauma to vascular walls and surrounding tissues (Fig. 4.26). Tissue trauma causes the release of tissue thromboplastin, a collection of

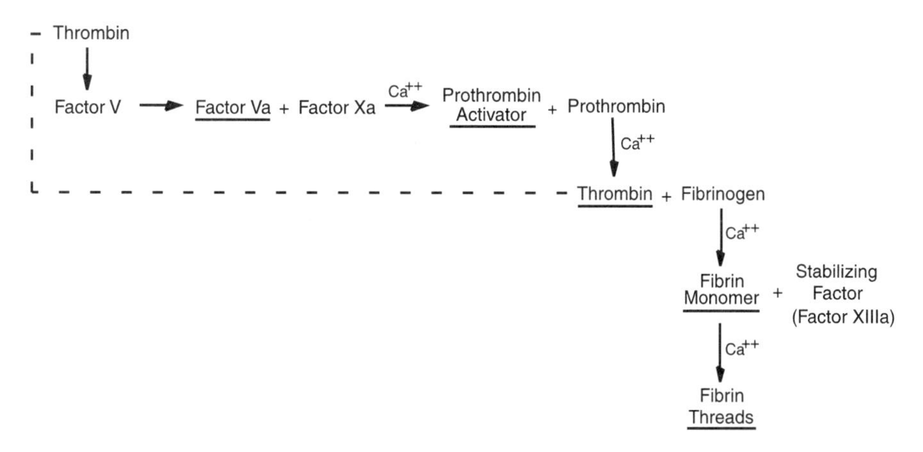

Figure 4.25. Contributions of thrombin to the production of fibrin threads. Through a series of enzymatic reactions described in the text, thrombin plays a key role in the formation of fibrin. Products of reactions are underlined to help make the progression of reactions clearer. A potential feedback loop is designated by the dashed line.

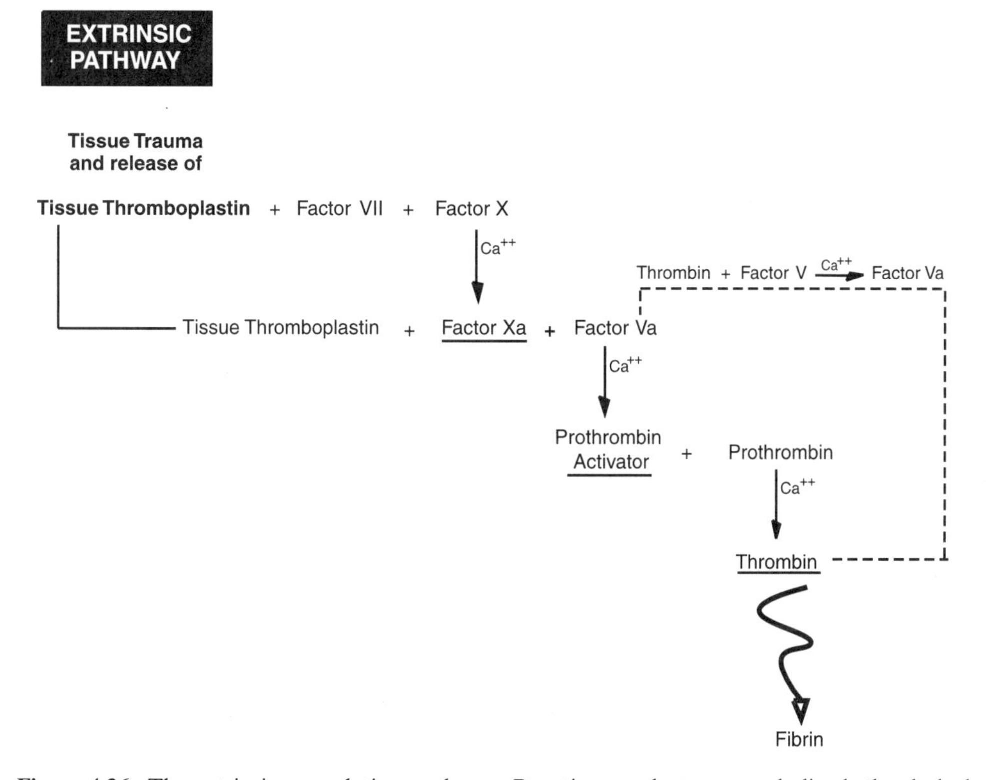

Figure 4.26. The extrinsic coagulation pathway. Reaction products are underlined; the dashed lines denote potential feedback loops. After the production of thrombin has been achieved, the reactions outlined in Fig. 4.24 occur.

several factors including phospholipids from damaged cell membranes as well as a lipoprotein complex. The lipoprotein complex can function as a *proteolytic* (protein digesting) enzyme. Tissue thromboplastin combines with and activates factor VII, and this complex is then able to enzymatically activate factor X. Factor Xa combines with phospholipids (from the tissue thromboplastin or from platelets) and with factor Va to form prothrombin activator. At this point in the extrinsic cascade, the same reactions diagrammed in Figure 4.27 occur, with the same positive feedback loop due to thrombin and with the same outcome: the formation of fibrin.

The *intrinsic pathway* is a coagulation cascade that begins with the exposure of blood to a foreign surface (i.e., a surface that is not the membrane of a normal, healthy endothelial cell) or with trauma to platelets within the blood (via, for example, interactions with a foreign surface). The intrinsic pathway (Fig. 4.27) can be initiated on blood contact with collagen in the subendothelium or on blood contact with a biomaterial. This is an important point to emphasize from a biomaterials perspective: Blood coagulation is automatically initiated upon implantation of a biomaterial or a device. In Chapter 5 we see that coagulation automatically triggers the inflammation phase of wound healing. Therefore, although the degree or severity of coagulation, inflammation, and subsequent events at the tissue-biomaterial interface depends on the surface properties of the biomaterial, every biomaterial will elicit biological responses from the body. To date, no biomaterial is inert.

To start the intrinsic cascade, factor XII binds to the foreign surface and is activated, starting a chain of reactions. Platelets that adhere to the foreign surface (which is defined as a surface other than the membrane of a healthy endothelial cell) or that are damaged release phospholipids from their membranes, as well as a compound called platelet factor 3. These platelet-derived compounds can participate in the intrinsic cascade at two points: during the activation of factor X and in the formation of prothrombin activator (just as the tissue thromboplastin was necessary for the formation of prothrombin activator in the extrinsic pathway). Note that the intrinsic pathway eventually works down to thrombin production and functions, as diagrammed in Figure 4.25. Because the intrinsic and extrinsic pathways join at the activation of factor X and culminate in the same thrombin-mediated mechanisms, the series of reactions from factor X to the formation of fibrin is often called the *common pathway* (Fig. 4.28).

Understanding the initial mechanisms of intrinsic pathway activation is crucial to understanding the difficulties in designing biomaterials for blood-contact applications. When a biomaterial contacts blood, a variety of circulating proteins quickly adsorb to the material surface. Two compounds that bind to surfaces quickly are high-molecular-weight kininogen (or HMWK) and factor XII (Fig. 4.29). Factor XI and prekallikrein circulate in the blood bound to HMWK, so when HMWK adsorbs to a surface, that process often brings prekallikrein and factor XI into the proximity of adsorbed factor XII. On adsorption, some of the factor XII is activated; this factor XIIa can then convert prekallikrein to

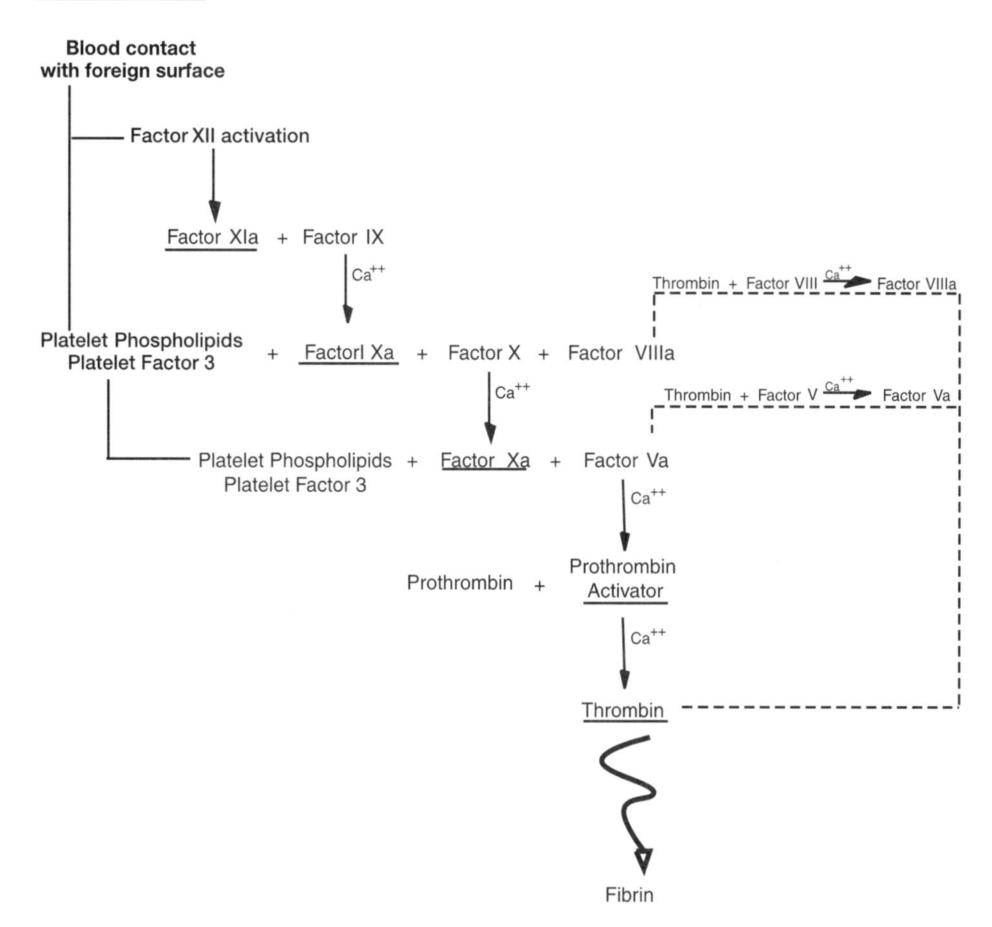

Figure 4.27. The intrinsic coagulation pathway. Reaction products are underlined; the dashed lines denote potential feedback loops. After the production of thrombin has been achieved, the reactions outlined in Fig. 4.24 occur.

kallikrein. Kallikrein can then reciprocally activate any factor XII that was not activated when it adsorbed to the surface. Factor XIIa then activates factor XI, and the intrinsic pathway continues. Interestingly, once the prekallikrein has been converted to kallikrein, the HMWK-kallikrein complex can desorb from the surface. This active kallikrein can now go on to enzymatically cleave small polypeptides called *kinins* from circulating serum proteins. The kinins (for example, *bradykinin*) cause vasodilation and increase vascular permeability, which are important to the inflammation stage of wound healing that follows coagulation (as outlined in the Introduction to this book). Kinins can also stimulate pain receptors and cause smooth muscle contraction, aiding in the vascular spasm response.

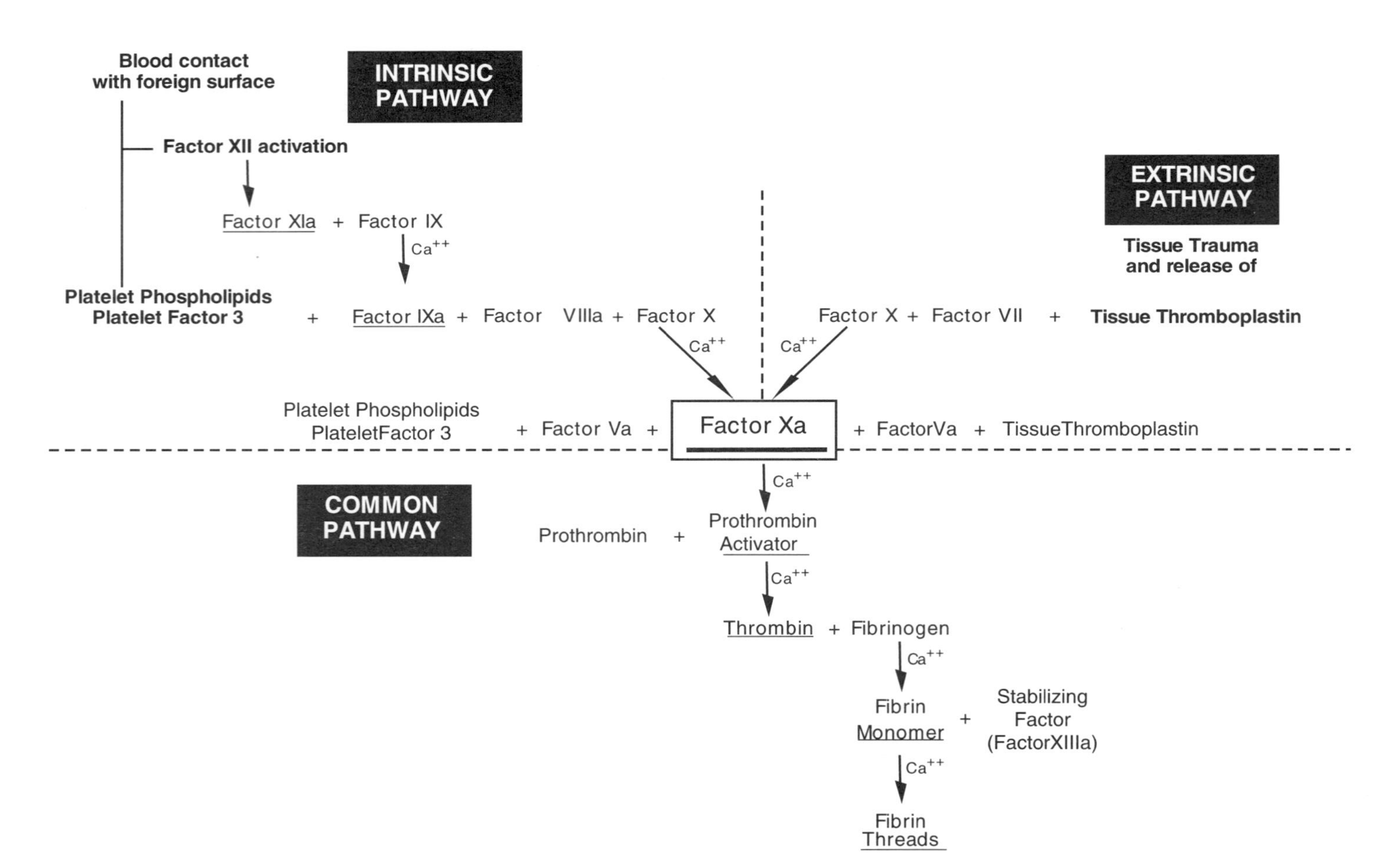

Figure 4.28. The coagulation cascades, with the common pathway delineated.

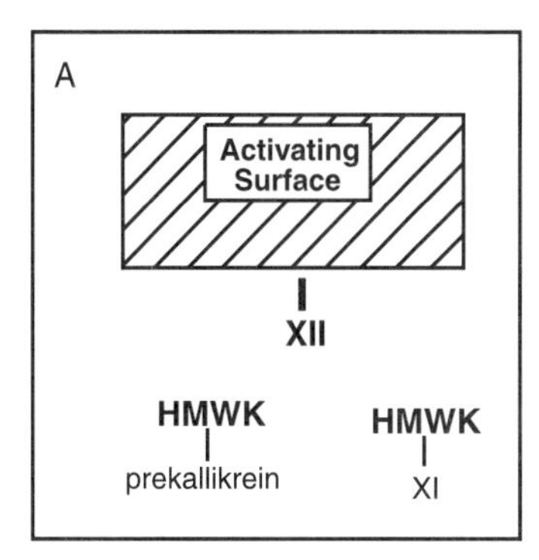

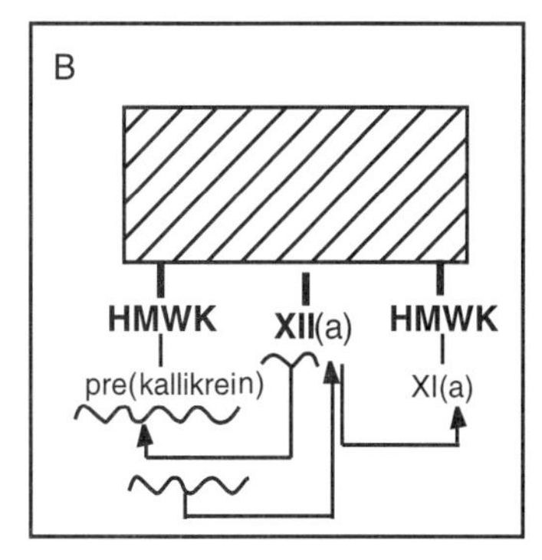

Figure 4.29. The initial steps of the intrinsic coagulation pathway. (**A**): factor XII adsorbs to a surface, followed by an HMWK-prekallikrein complex and an HMWK-factor XI complex. (**B**): The factor XII that was activated on adsorption acts on the prekallikrein to form kallikrein. The kallikrein in turn then activates any remaining unactivated factor XII. The activated factor XII then activates the factor XI, and the intrinsic pathway has begun.

4.5.2 Control Points

The coagulation cascades can be viewed as a complex engineering system, with potential feedback and feedforward loops. Physicians and biomedical engineers may need to exert varying degrees of control over the coagulation process for a number of reasons. For example, it is undesirable for blood to coagulate when it is being donated for transfusion into another person, so blood collection bags are designed to inhibit coagulation. Some diagnostic tests measure the speed of blood clotting under various conditions, so for these applications delicate control over coagulation is needed. Considering the coagulation cascades as a system allows us to identify control points, or steps within the cascades that are vital to controlling the process.

Calcium. Calcium ions are required for all reactions in the coagulation cascades except for the activation of factor XII and factor XI. In the absence of calcium ions, blood will not clot. Precipitating or chelating calcium (with, for example, ethylenediaminetetraacetic acid, or EDTA) is one way to allow storage of blood outside the body without inducing coagulation.

Thrombin. Note that thrombin is a key element in the common pathway and in feedback loops. It is then logical that controlling the availability and reactivity of thrombin would allow a good deal of control over the end stages of coagulation.

Platelet contributions. Platelets adhere quickly to foreign (nonendothelial) surfaces, initiating an outbreak of platelet aggregation. Understanding how platelets adhere to surfaces may allow the design of biomaterial surfaces that could be largely "ignored" by platelets. Moreover, platelet phospholipids (either released from the membranes of disrupted platelets or displayed on the surface of adherent and aggregating platelets) play roles in both the extrinsic and intrinsic pathways. In many ways, the platelet membrane can be viewed as a surface catalyst for coagulation.

Clotting factors. Factor X is the link between the extrinsic and intrinsic pathways, making this factor a key control point. Deficiency of factor VIII is the cause of hemophilia A, a disease caused by a recessive defect on the X chromosome, perhaps most famously identified in royal families of Europe related to Queen Victoria of England. Patients with hemophilia A experience excessive bleeding from mild trauma, sometimes only subsiding on an injection of factor VIII.

4.6 ANTICOAGULANTS AND FIBRINOLYSIS

Biomedical engineers seeking to understand and control coagulation may benefit from strategies similar to some of the biological mechanisms that keep the coagulation reactions in check. One nonspecific biological control mechanism is blood flow: The passing blood volume dilutes coagulants and removes them from the local area of injury. Activated blood clotting factors are removed from the circulation in the liver. Another somewhat nonspecific control mechanism is the fact that several steps within the coagulation cascade are favored only when catalyzed on the surface of activated platelets or at a local stimulus. For example, activation of factors XII and XI is surface mediated and provides a localized (immobilized on the surface) stimulus for coagulation to occur. Additionally, factors IXa and VIIIa form a complex on the phospholipids typically expressed on the surface of activated platelets; this complex activates factor X. The conversion of prothrombin to thrombin is also catalyzed by the surface of activated platelets. The membranes of activated platelets thus help localize the coagulation reactions to the site of injury as well as helping the reactions to occur at a significant rate. Without platelet phospholipids, coagulation reactions occur very slowly, if at all.

There are a number of naturally occurring and specific anticoagulants. The endothelium itself prevents coagulation by its surface characteristics. The smooth and "slimy" nature of the glycocalyx mucopolysaccharide layer on the blood-contacting surface of the endothelium repels clotting factors and adhering platelets. An endothelial membrane protein called thrombomodulin binds thrombin, making it unavailable for further coagulation cascades. Additionally, the thrombomodulin-thrombin complex can activate a circulating, vitamin K-dependent plasma protein known as protein C, which then inactivates factor Va and factor VIIIa. Protein S, another vitamin K-dependent plasma protein, also inactivates factor Va and factor VIIIa. Deficiencies of protein C or protein S can lead to excessive clotting and the formation of blood clots that block the flow in arteries and veins.

Paradoxically, one of the most powerful naturally occurring anticoagulants is the fibrin that is formed during the coagulation process. Approximately 85–90% of the thrombin formed during coagulation adsorbs to the newly polymerized fibrin threads. This helps localize the coagulation reaction to the site of injury, rather than allowing the coagulation to spread throughout the blood system. Antithrombin III, a serum protein, binds and inactivates much of the rest of the nonadsorbed thrombin within 20 min. When antithrombin III (ATIII) is complexed with heparin, a polysaccharide that has little or no anticoagulant activity on its own, the effectiveness of ATIII at removing thrombin from the blood is increased from a hundred to a thousand times. The heparin-ATIII complex also is able to inactivate factors IXa, Xa, XIa, and XIIa, making heparin-ATIII a powerful anticoagulant able to slow or stop the intrinsic, extrinsic, and common pathways. Usually there is an excess of ATIII in human blood; the amount of heparin present is usually the limiting factor that determines the efficacy of ATIII-mediated thrombin removal. Providing extra heparin (intravenously) to a patient is a common clinical anticoagulant strategy.

The anticoagulants discussed above function to localize and control blood clotting. Once a clot has formed, does it passively degrade over time or is it actively removed? Many people think that blood clots simply break down as they age and eventually fall apart, reasoning from personal knowledge that minor skin wounds take a few days to heal. Actually, blood clots are actively removed by an enzymatic digestion process that includes a typical time delay of 24–48 h, allowing the surrounding tissues some time to heal before the stabilizing influence of the clot is removed. During coagulation, *plasminogen*, a circulating plasma protein, is trapped in the forming clot (along with a number of other plasma proteins). Injured tissues release a complex called *tissue plasminogen activator* which, over the course of 1–2 days, converts the plasminogen into *plasmin*. Plasmin, a proteolytic enzyme similar to trypsin, digests the fibrin threads of the clot as well as other procoagulants such as fibrinogen, prothrombin, and factors V, VIII, and XII. Small amounts of plasmin are activated in the bloodstream all the time. Because plasmin removes both end-stage and intermediate-stage coagulation products, circulating active plasmin could significantly decrease the efficacy of the coagulation process if it were not controlled by interactions with a serum protein called α_2-antiplasmin, which inactivates plasmin. At a coagulation site, the local amount of plasmin has to rise above a critical level suppressible by passing α_2-antiplasmin to be an effective *fibrinolytic* (fibrin destroying) agent.

4.7 BIOMATERIALS, DEVICES, AND THROMBOSIS

The normal, healthy endothelium does not induce coagulation; it is the perfect nonthrombogenic blood-contacting material. Major advances have been made in designing blood-contacting biomaterials and devices during recent decades, and there is a good deal of published literature on efforts to control blood-biomaterial interactions. This section is not a comprehensive review of the research field of cardiovascular materials but is instead intended to summarize a few major concepts and questions that students should be familiar with. Specifically, we attempt to briefly address these questions: What types of events tend to occur at the blood-biomaterial interface? How could these events affect a patient or a device? What are some general ideas for influencing the blood-compatibility of biomaterials? What are some major areas of ongoing research in this field?

In an excellent reference book titled *Biomaterials Science: An Introduction to Materials in Medicine* (see Section 4.9 for full bibliographic information), authors Didisheim and Watson have listed blood-material interactions in two categories that are paraphrased here: those that may affect the function of the material (or device) and those that could be detrimental to a patient. When considering the function of a blood-contacting material or device, scientists and engineers should be aware of the potential effects of adsorption of blood components (such as proteins, lipids, or calcium) onto (or deposition into) the surface of the material, as well as adhesion of blood cells or components of blood cells onto the surface of the material. For implanted devices, tissue growth around the device and potentially even the production of a new intima on the blood-contacting portion of the device should be considered as possibilities. When considering outcomes of blood-material interaction that could be detrimental to a patient, Didisheim and Watson note that a number of factors should be considered, including: the processes of coagulation and fibrinolysis, as well as activation of the immune system; formation of *thrombi*, or clots, on the surface of the material; production of *emboli*, or clots that break away from a surface and travel through the circulatory system, potentially blocking blood vessels elsewhere; and injury to blood cells that could lead to low levels of blood-borne cells and subsequent clinical problems. For implanted devices, *intimal hyperplasia* or overgrowth of the intimal tissue near or on a device can significantly reduce flow in blood vessels.

Many scientists and engineers have tried to produce more blood-compatible biomaterials by mimicking select properties of the endothelium. For example, very smooth materials are typically used in blood-contact applications. Specific endothelial membrane components can also be imitated, such as surfaces modified to contain negative charges, or with "slippery" components of an imitation glycocalyx. Various hydrophilic biomaterials have been shown to reduce platelet adhesion and thrombus formation, as compared to more hydrophobic materials. Many research groups have incorporated anticoagulants, such as heparin, on blood-contacting surfaces. The anticoagulant can be immobilized on the biomaterial surface either for controlled release or for a longer-term presence; alternatively, an anticoagulant may be incorporated into the chemical structure of a polymer, becoming an integral part of the biomaterial.

Many current biomaterials design strategies show promise, and new developments in blood-biomaterial compatibility seem to occur every year. However, there are still major clinical challenges to be solved. Three of these challenges are summarized here.

1. Vascular Grafts. Replacing narrowed or occluded portions of larger arteries is becoming a fairly common medical procedure. Currently, the best vascular graft performance is given by *autografts* (tissue taken from one site on a patient and transplanted to another site on the same patient). Often the saphenous vein is used as a vascular autograft. The main mechanisms of vascular autograft failure are thrombosis, emboli production, and *intimal hyperplasia*. Intimal hyperplasia is an overgrowth of the (normally quiescent) smooth muscle cells causing the intima to bulge into the blood vessel area, often observed at *anastomotic sites* (the sites where the graft was sutured into the original blood vessel). Intimal hyperplasia can significantly decrease or even block blood flow through the graft. It is thought that both mechanical and chemical factors can contribute to intimal hyperplasia. Cells in arteries experience higher degrees and larger variations in fluid shear due to blood flow than do cells in veins, so the cells in the vein grafts may be responding to the new mechanical environment by overgrowing. The anastomotic sites may be focal points for increased mechanical strain on the surrounding tissues, as the tissues strain against the sutures. These strains may also help to cause tissue overgrowth. Endothelial cell injury and death during the vein harvest and transplantation may stimulate platelet activation and coagulation; some of the products released by platelets during aggregation (for example, platelet-derived growth factor) may stimulate the proliferation of smooth muscle cells. Intimal hyperplasia is discussed further in Chapter 10.

The use of synthetic vascular grafts is necessary when (because of, for example, size mismatches or disease state) a patient's veins cannot be used as autografts. One popular synthetic vascular graft material that has shown good clinical results is knitted Dacron, widely used for medium- and large-diameter graft sites. These grafts may be exposed to a patient's blood to allow "preclotting" before implantation, to prevent excessive blood leakage through the fabric. Another popular synthetic vascular graft material is expanded poly(tetrafluoroethylene), which exhibits very low incidences of thrombosis or hyperplasia—as long as the inner diameter of the graft is greater than approximately 4 mm. Smaller-diameter vascular grafts, often needed when replacing vessels below the knee, still show unsatisfactory performance (in general) no matter what biomaterial is used. This may be caused by prolonged blood-biomaterial contact due to the reduced flow from the small diameter or by the relatively high surface-to-volume ratio of these grafts, which may increase the activation of surface-mediated coagulation.

2. Blood oxygenators. The cardiopulmonary bypass, or heart-lung, machine pumps deoxygenated blood from a patient to an oxygenator and returns oxygenated blood to the arterial circulation, bypassing the lungs. These machines are a masterpiece of medical engineering, and they make possible open-heart surgery, oxygenation therapy for patients in severe respiratory distress, and keeping patients alive after severe heart/lung trauma. However, blood in a cardiopulmonary bypass machine must traverse a wide variety of foreign biomaterial surfaces (tubing, reservoirs, fibers, plates, etc.), survive fluid shear and mechanical stresses from the pumping action, be maintained at an appropriate temperature without over- or underheating, and, finally, be oxygenated without significant damage to the red blood cells. Currently, there are two main types of blood oxygenators in use: bubble oxygenators, in which oxygen gas is mixed directly with the blood, and membrane oxygenators, in which oxygen diffuses across a membrane for uptake by red blood cells. Blood cell trauma tends to occur at the air-liquid interfaces in bubble oxygenators, so great care must be taken to minimize foaming in this application. Less blood trauma is caused by (the more expensive and complex) membrane oxygenators, but varying degrees of *hemolysis* (erythrocyte death), platelet loss, and circulating coagulation factor depletion can still occur. Systemic anticoagulants must be used during cardiopulmonary bypass, which may cause undesirable side effects and disruption of hemostasis. The development of increasingly blood-compatible biomaterials for use in the blood-contacting surface areas and oxygenation membranes of cardiopulmonary bypass machines will make this important medical device even more useful.

3. Heart Valves. Replacing a diseased or defective heart valve with an artificial valve can significantly improve a patient's quality of life. Heart valves made from synthetic materials typically utilize a caged ball or a tilting disk to periodically permit blood flow through the valve. Synthetic heart valves have a history of innovation and continual redesign because of various materials processing and fabrication issues, as well as the severe environment experienced by these prostheses (highly repetitive movement causing mechanical fatigue, chemical interactions with the blood, etc.). Although a variety of materials have been used in synthetic heart valves, thromboembolism may still occur because of biomaterial surface-activated coagulation. With the goals of reducing thromboembolism, enhancing integration of the artificial valve into the surrounding tissues, and reproducing the natural "leaflet" closing mechanism of healthy heart valves, tissue heart valves have been and are being developed.

Tissue valves may be *allografts*, or a transplant from a member of the same species (in this case, from a human donor), or *xenografts*, transplants from a member of a different species (for heart valves, porcine or bovine tissues are often used). Allograft heart valves can be obtained at autopsy and cryopreserved for subsequent use. However, these allografts are rarely used because of the difficulties in obtaining the necessary valve size and limited availability in general. Xenograft heart valves, generally porcine valves or valves constructed from bovine pericardium, are an area of much current research. These valves can be less prone to thromboembolism than synthetic valves, but they present new problems. The tissue must be chemically treated so that the patient doesn't have an immune reaction to the foreign tissue (discussed in Chapter 6). Treatment with glutaraldehyde appears to reduce the potential for an immune response satisfactorily, but glutaraldehyde also chemically cross-links the collagen in the tissues. These cross-links stiffen the tissue leaflets, changing the mechanical properties of the tissue, even though the flexibility of the tissue leaflets is crucial to the long-term performance of the valve. Tissue degeneration, calcification and stiffening, and subsequent fatigue-related failure are all issues to be explored and countered in the continuing development of xenograft heart valves. See Section 4.9, Bibliography/Suggested Reading, for reference information on a recently published comprehensive review of issues surrounding the continued development of tissue heart valves.

Each of the three example devices briefly introduced above illustrates that what could be loosely termed the "mechanical environment" (fluid flow and shear, material compliance, mechanical fatigue, etc.) of a blood-contacting device is an important factor in determining ultimate device success or failure. From understanding how coagulation is controlled in vivo, it becomes clear that the "chemical environment" (utilizing anticoagulants to strategically target control points in the coagulation process) of a blood-contacting material might also be manipulated effectively to achieve desired clinical outcomes. However, coagulation is only the first phase of the wound healing process. The second stage, inflammation, poses a number of new challenges for biomedical engineers seeking to control tissue-biomaterial interactions.

4.8 SUMMARY

- All of the cells that circulate in the bloodstream are ultimately derived from pluripotent hematopoietic stem cells in the bone marrow.
- Although erythrocytes play minimal roles in wound healing and in blood-biomaterial interactions, the production and function of these cells can be used to introduce and demonstrate fundamental biological and engineering concepts, including:
  - Sequential differentiation to develop specialized structures and corresponding functions of cells;
  - Fluid flow characteristics of blood as a function of shear rate and hematocrit;
  - Osmolarity and related cell membrane distension;
  - How small changes in biomolecule chemical structure can cause large-order functional changes, such as sickle cell.
- The residual megakaryocyte components in platelets allow these cellular fragments to be functionally active and responsive to the extracellular environment. Platelets play major roles in coagulation, including the formation of the platelet plug, contributions from the arachidonic acid cascade, and membrane-catalyzed coagulation reactions.

- The extrinsic coagulation cascade begins with trauma to vascular walls and surrounding tissues and the intrinsic coagulation cascade is initiated by the exposure of blood to a foreign surface, but both pathways culminate in the same thrombin-mediated mechanisms (common pathway). Considering the coagulation cascades as an interconnected system allows the identification of "control points" for the process as a whole (i.e., calcium, thrombin, factor X).
- Because blood contact with a foreign surface (a surface other than a healthy endothelial cell membrane) initiates the intrinsic pathway, and because coagulation automatically triggers the inflammatory stage of the wound healing process, every biomaterial will elicit biological responses from the body. To date, no biomaterial is perfectly inert.
- There are both nonspecific and specific naturally occurring anticoagulants. Endothelial cells provide some anticoagulants, whereas others are nonendothelial in origin (for example, thrombin). Understanding how anticoagulation and fibrinolysis occur naturally can provide ideas for controlling these processes in medical devices and situations that require good blood-biomaterial compatibility.
- The "mechanical environment" (fluid flow and shear, material compliance, mechanical fatigue, etc.) of a blood-contacting device plays an important role in the long-term clinical utility of the device.

4.9 BIBLIOGRAPHY/SUGGESTED READING

Didisheim, P. and Watson, J.T., "Cardiovascular Applications" in *Biomaterials Science: An Introduction to Materials in Medicine*, (Ratner, B.D., Hoffman, A.S., Schoen, F.J. and Lemons, J.E., eds.) Academic Press, New York, NY (1996), pp. 283–297.

Fung, Y.C., "Chapter 3: The Flow Properties of Blood," in *Biomechanics: Mechanical Properties of Living Tissues*, second edition, Springer-Verlag, New York, NY (1993), pp. 66–108.

Guyton, A.C., "Chapter 36: Hemostasis and Blood Coagulation," in *Textbook of Medical Physiology*, eighth edition, W.B. Saunders Company, Harcourt Brace Jovanovich, Inc, Philadelphia, PA (1991), pp. 390–399.

Hayhoe, F.G.J. and Flemans, R.J., *Color Atlas of Hematological Cytology*, third edition, Mosby-Year Book, Inc., St. Louis, MO (1992).

Junquiera, L.C., Carneiro, J., and Kelley, R.O., "Chapter 12: Blood Cells," in *Basic Histology*, eighth edition, Appleton & Lange, Norwalk, CT (1995), pp. 218–222.

Schoen, F.J. and Levy, R.J., "Tissue Heart Valves: Current Challenges and Future Research Perspectives," *Journal of Biomedical Materials Research*, 47: 439–465, 1999.

4.10 STUDY QUESTIONS

1. Describe the constituents of blood.
2. Explain where hematopoietic cells come from, identifying important intermediate stem cells and terminally differentiated cells.
3. Make a series of sketches to explain the process of erythrocyte formation, including the contribution of erythropoietin and the concepts of cellular differentiation, replication, and nuclear expulsion.
4. Draw a diagram summarizing the effects of shear rate and hematocrit on the viscosity of blood. If the blood contained sickled cells, how would you expect your diagram to change? If you varied the osmolarity of blood containing normal erythrocytes, how would you expect the viscosity to change?
5. Explain the process of platelet formation, including a description of how the platelet and the erythrocyte differ in function even though both lack a nucleus.
6. List potential mechanisms of platelet adhesion to collagen or a biomaterial surface.
7. Describe the physiological relevance and sequence of events involved in forming the "platelet plug."
8. Diagram how thrombin contributes to the formation of fibrin.
9. Explain, in detail, why the degree of trauma in a wound is related to the degree of coagulation.
10. Create your own hand-drawn chart that diagrams the processes of coagulation, including the contributions of platelets. Don't limit yourself to one 8.5×11-in. sheet of paper—it will probably take more space. Use color to highlight what you consider to be key steps and control points within the processes. Now add notations to the chart for the actions of anticoagulants and for the process of fibrinolysis.
11. Explain what intimal hyperplasia is.

4.11 DISCOVERY ACTIVITIES

1. How could you design a biomaterial surface that would control the intrinsic pathway of coagulation? How about the common pathway? Or platelet contributions? Justify your design strategies.
2. Defend the statement that "no biomaterial is inert."
3. Define autograft, allograft, and xenograft; research and discuss clinical issues associated with these graft types in the applications of heart valves, vascular grafts, and bone tissue.
4. Professional cyclists may undergo blood tests to confirm their hematocrit before

competing in high-profile races such as the Tour de France. Why? What is the maximum allowable hematocrit level for cyclists associated with the United States Cycling Federation? How does this level compare to typical physiological levels for men and women?

5. Find a company that manufactures and sells blood oxygenator devices, vascular grafts, or heart valves. What biomaterials are used by the company in its product? What are some of the major benefits of the company's product as compared with other products on the market? How does the company test blood-biomaterial compatibility for its product?

6. Recent reports in the scientific literature indicate that many types of stem cells may be more totipotent than originally thought. Find an article in the popular press and a paper in a scientific journal about stem cells from one type of tissue being used to produce cells that appear to be from another type of tissue. Do the two articles present the same fundamental conclusions? What types of methods were used to encourage the stem cells to produce their unusual progeny? How were the progeny cells characterized—what makes the researchers think that the stem cells truly produced cells of another tissue? What ethical issues are associated with the procurement and study of human stem cells?

7. Some researchers and companies are working on blood substitutes—solutions that do not depend on viable erythrocytes to carry oxygen. What types of blood substitutes are currently under development? What scientific challenges have yet to be met?

5

Inflammation and Infection

5.1 INTRODUCTION

The second general stage of the wound healing process is inflammation. Many of the hematopoietically derived cells that were not discussed in detail in Chapter 4 play major roles in the inflammatory process: removing cellular and tissue debris from the site of injury; destroying any foreign material, bacteria, or microorganisms present in the wound; and secreting chemicals that attract other cell types to the wound to produce new tissue. Inflammation is a normal and necessary part of the wound healing process, and all implanted biomaterials are exposed to some degree of the inflammatory process as the implantation site heals.

5.2 HISTORICAL OBSERVATIONS: INFLAMMATION AND INFECTION

Medical personnel have been studying inflammation and infection for a long time. A word that can be interpreted to mean "inflammation" (Fig. 5.1) was used in connection with descriptions of wounds in the hieroglyphs of the Smith Papyrus (a document written in Egypt around 1650 B.C. and originating from another document that may have been 1,000 years older) [1, 2]. The sign at the far right of Figure 5.1 was evidently not meant to be pronounced but was to give an overall idea or impression of the preceding word. The unspoken sign in Figure 5.1 is meant to depict a flaming coal with smoke rising up and then falling down again, giving the idea of a "hot thing," and the ancient Greeks (400–450 B.C.) had a word for inflammation that roughly translates into English as "the burning thing." [1] In the historical periods before the widespread application of current medical hygiene standards (much less antibiotics), inflammation and infection often occurred simultaneously in wounds. However, these are two separate processes, and they should be differentiated.

Infection is the result of microorganisms such as bacteria, fungi, or viruses colonizing tissues. A sting from a bee or wasp causes inflammation, but not necessarily infection. A mild burn can cause skin inflammation without the bacteria

Figure 5.1. Hieroglyphs reading "shememet," thought to designate inflammation. The vowel "e" is given as a guess to fill the blanks between the consonants "sh-m-m-t". Redrawn from page 6 of G.B. Ryan and G. Majno, *Inflammation*, the Upjohn Company (1980), where the Smith Papyrus (1650 B.C.) is cited as the original source of this figure.

growing in the burned tissue, so *inflammation does not imply infection*. In contrast, the body's response to infectious agents often includes typical inflammatory symptoms such as painful, swollen, warm areas of tissue. A key distinguishing feature of infection, but not of inflammation, is that infected wounds often *suppurate* (produce pus). Historically, therefore, most wounds healed only after suppuration, leading people to (mistakenly) believe that suppuration was a normal and necessary step of wound healing. People perceived suppuration to indicate that a harmful fluid, probably derived from decaying blood, needed to drain from wounds for healing to continue. Wounds were treated by reopening to encourage drainage, and people were bled to get the harmful "decaying blood" out of them. Bloodletting was even thought to prevent inflammation (which occasionally it did, by killing the patient).

In the first century A.D. a Roman named Celsus [2] gave an excellent definition of inflammation, in what has become a famous piece of medical Latin: "*Rubor et tumor cum calore et dolore*," or "Redness and swelling with heat and pain." These are known as the four *cardinal signs* of inflammation. A fifth cardinal sign, *functio laesa* or "disturbed function," is sometimes added to emphasize the fact that inflamed organs do not function properly. Current knowledge of biology and physiology does not negate the description of inflammation provided by the cardinal signs but rather allows us to explain the underlying causes of these symptoms. Inflammation (not infection) is a normal, necessary stage of wound healing (Fig. 0.2), which follows the process of coagulation. A good starting point in understanding inflammation is to identify and characterize the functions of the involved cells.

5.3 NONLYMPHATIC LEUKOCYTES

In Chapter 4, we differentiated between lymphatic and nonlymphatic leukocytes. The lymphatic leukocytes participate in the body's immune response to foreign (non-self) materials, chemicals, or cells; they are discussed in Chapter 6. The nonlymphatic leukocytes are the focus of this section, and their lineage is shown in Figure 5.2. The terminally differentiated cells (Figs. 5.3 and 5.4) are

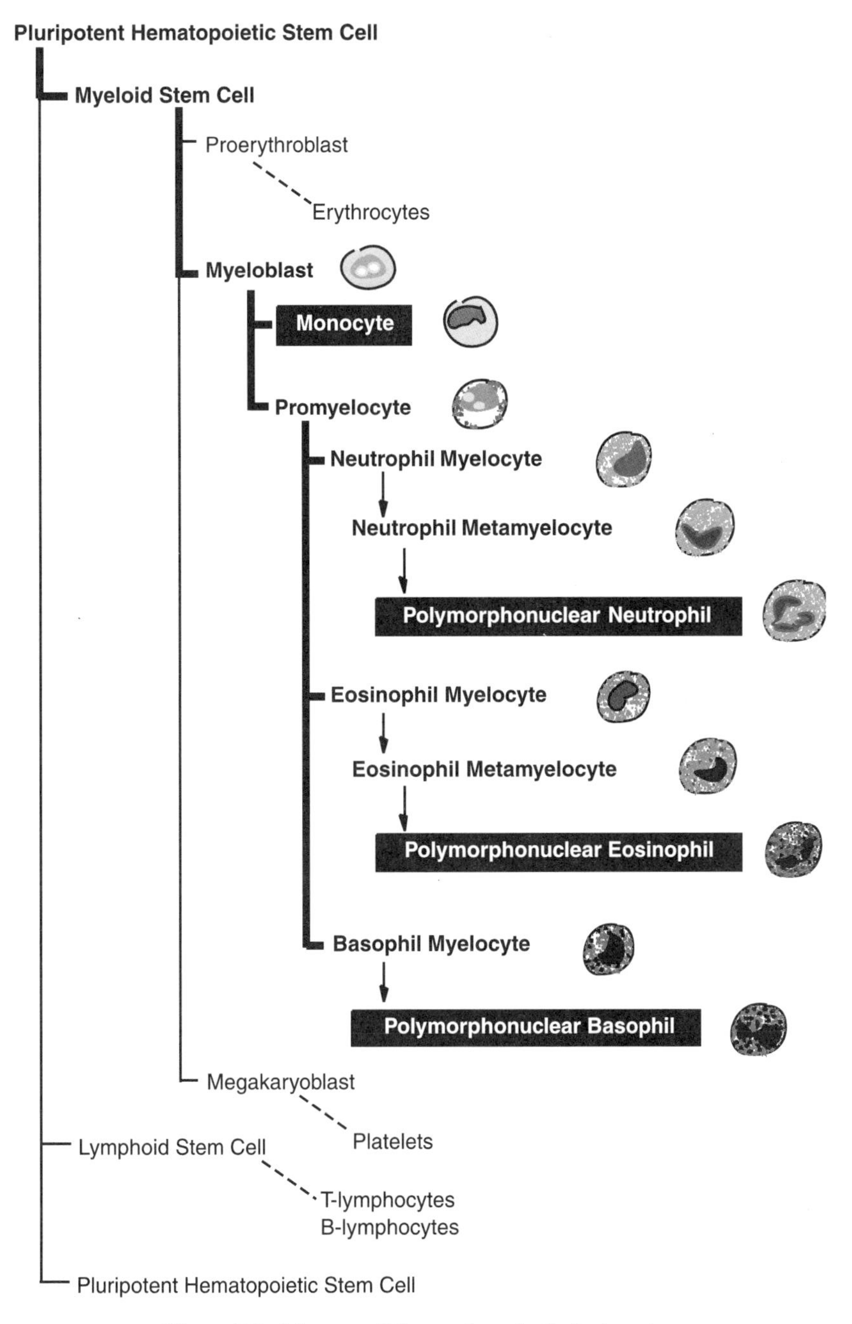

Figure 5.2. Lineage of the nonlymphatic leukocytes.

classified by the presence of intracellular granules (present in neutrophils, eosinophils, and basophils; the *granulocytes*), the lobed shape of the nuclei (*polymorphonuclear*), and the types of histologic stains that are readily bound by the intracellular components of each cell. All of the nonlymphatic leukocytes are

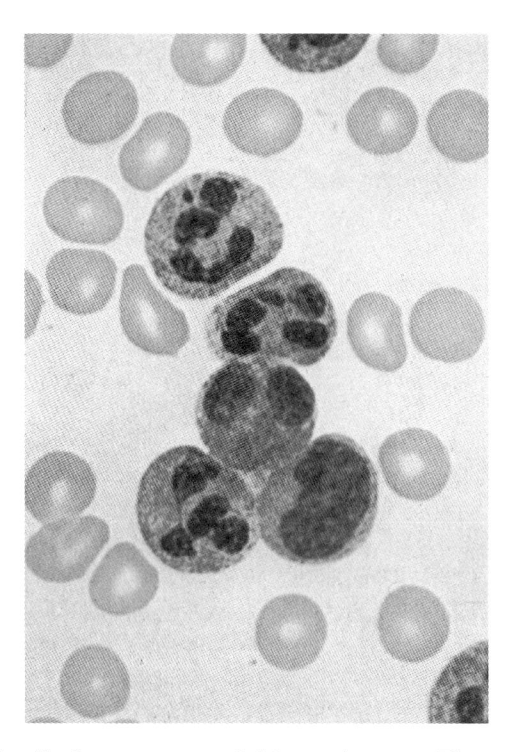

Figure 5.3. Micrograph of leukocytes surrounded by erythrocytes. The top three cells in the center of the picture are, from top to bottom, two polymorphonuclear neutrophils and one polymorphonuclear eosinophil. The bottom two cells in the center of the picture are another polymorphonuclear neutrophil (left) and a monocyte (right). The other cells visible in their entirety are erythrocytes. Figure reproduced from F.G.J. Hayhoe and R.J. Flemans, *Color Atlas of Hematological Cytology*, page 98, © Mosby-Year Book, Inc., 1992. Reprinted by permission of W.B. Saunders Company/Mosby/Churchill Livingstone.

formed in the bone marrow, enter and circulate through the bloodstream, and then leave the bloodstream and migrate through tissues. Their fundamental purpose is to kill, consume, or destroy "foreign objects," which may include particulate debris from biomaterials, bacteria, pieces of damaged tissue, or dead cells. Approximately 6,000–10,000 leukocytes are present per microliter (or mm^3) of blood in a normal adult (Table 5.1).

Approximately three times as many granulocytes are stored in the bone marrow as are circulating in the blood at any given time. On release from the bone marrow, granulocytes tend to circulate through the bloodstream for approximately 4–8 hours, followed by a typical life span of 4–5 days in the tissues. In the case of serious infections, the life span of granulocytes can be dramatically shortened to as little as a few hours, because the cells quickly migrate to the infection site and die in the process of trying to destroy the invading *pathogens* (disease-causing bacteria, viruses, fungi, etc.).

Basophils are similar in function (but not in origin) to the *mast cells* typically

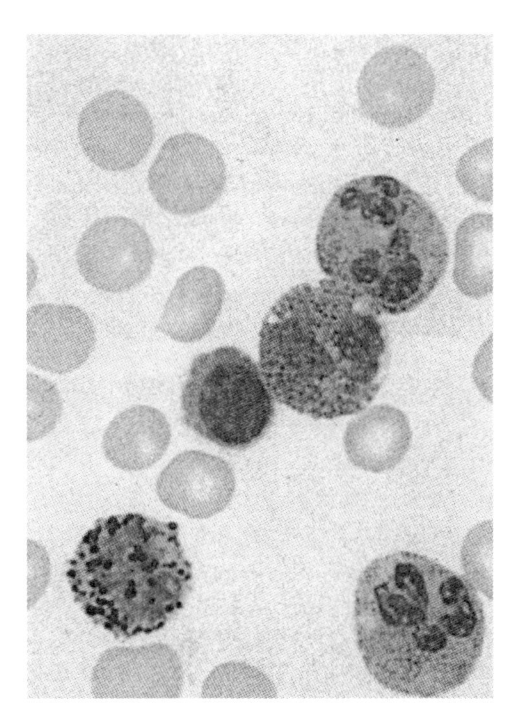

Figure 5.4. Micrograph of leukocytes and a lymphocyte. The nonerythrocyte cells in this image are, from top to bottom of the image, a neutrophil, an eosinophil, a lymphocyte, a basophil, and another neutrophil. Figure reproduced from F.G.J. Hayhoe and R.J. Flemans, *Color Atlas of Hematological Cytology*, page 108, © Mosby-Year Book, Inc., 1992. Reprinted by permission of W.B. Saunders Company/Mosby/Churchill Livingstone.

found near (outside of) capillaries. Both release heparin into the blood; both are also involved in a specific kind (IgE mediated; discussed in Chapter 6) of "allergic" immune response, which includes the rupture of these cells and the release of chemicals that cause systemic and local allergy symptoms. Both basophils and mast cells can release *histamine*, *bradykinin*, and *serotonin*.

TABLE 5.1. Approximate Percentages of Leukocytes in Adult Human Blood

Leukocyte Type	Concentration in Blood (%)
PMN neutrophils	62
Lymphocytes	30
Monocytes	5.3
PMN eosinophils	2.3
PMN basophils	0.4

Data taken from p. 366 of A.C. Guyton, *Textbook of Medical Physiology*, eighth edition, W.B. Saunders Company, Harcourt Brace Jovanovich, Inc., 1991.

Eosinophils fight off parasitic infections by attaching themselves to parasites and releasing lethal chemical compounds, such as enzymes, superoxides, and a larvacidal protein called major basic protein. Eosinophils also consume antibody-antigen complexes (discussed in Chapter 6).

Neutrophils are mature cells when released from the bone marrow into the circulatory system. The main function of neutrophils is *phagocytosis* ("cellular eating") or consumption of foreign particles or objects.

Monocytes are immature cells until they leave the bloodstream and migrate into tissues, at which time these cells swell as much as five times—up to diameters of 80–100 μm—and gain a granular appearance. These cells are now *macrophages*, and their main function is also to phagocytose foreign objects or particles. Macrophages typically live from 2 to 4 months in tissues. Some macrophages, on entering the tissues, remain there as a "first line of defense" against infection and can survive for time periods up to years. Tissues that are logical points of entry for infectious agents or particles tend to house these tissue macrophages—skin and subcutaneous tissues, lung alveoli, the liver sinuses that filter blood from the gastrointestinal tract, and the lymph nodes.

Macrophages can phagocytose greater quantities of particles or bacteria than can neutrophils. If macrophages ingest small amounts of particulate debris that resist degradation, these particles can remain sequestered in tissue macrophages until the death of the cell, when the particles can be rephagocytosed by another tissue macrophage. When confronted with the long-term presence of particles or objects too large to phagocytose (for example, a splinter or an implant), macrophages can fuse together to form large, multinucleated *foreign body giant cells* (Fig. 5.5). These large cells can phagocytose larger particles than can individual macrophages, and they have a life span of only a few days.

5.4 INFLAMMATION AND LEUKOCYTE FUNCTIONS

On injury, the coagulation process begins almost immediately and chemically stimulates the inflammation process to begin as well (discussed in Section 5.5). Shortly thereafter, the body's first cellular line of defense against any foreign objects or microorganisms that may be in the wound are activated: the tissue macrophages (described in Section 5.3) located near the injury. These cells migrate through tissue to the site of injury. How do they know which way to move?

5.4.1 Chemotaxis and Cell Migration

Chemotaxis is the term for cellular migration along a chemical gradient. Some receptors on the cell membrane are able to bind chemicals in the surrounding environment, and increased engagement of select receptors on one side of a cell

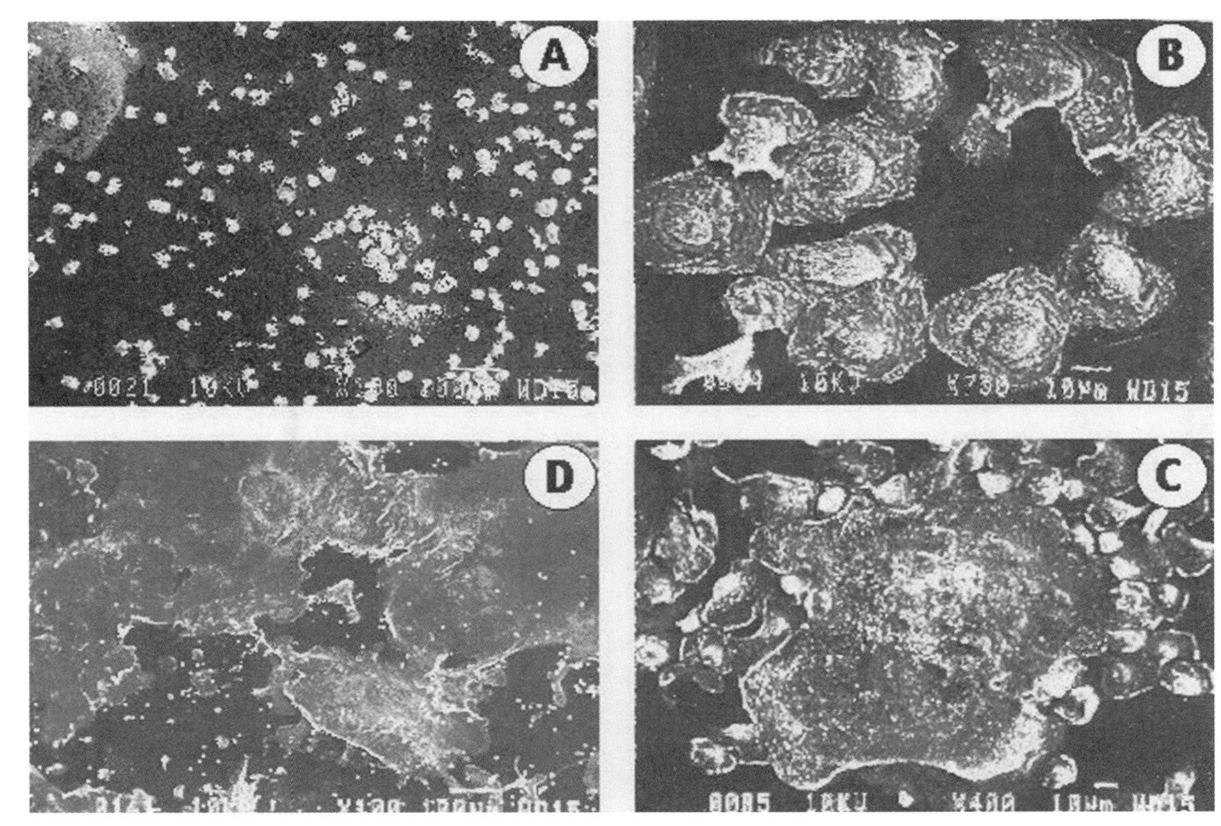

Figure 5.5. Scanning electron micrographs of the evolution of foreign body giant cells. **A** is a micrograph of monocytes; **B** is a micrograph of macrophages. Note the difference in scale between **A** (white scale bar = 100 μm) and **B** (white scale bar = 10 μm). In **C**, macrophages have merged together to form a multinucleated foreign body giant cell. In **D**, a number of foreign body giant cells have assembled to cover a biomaterial surface. Note also the difference in scale between **C** (white scale bar = 10 μm) and D (white scale bar = 100 μm). Figure courtesy of M. Shive, W. Brodbeck and J. Anderson, Case Western Reserve University.

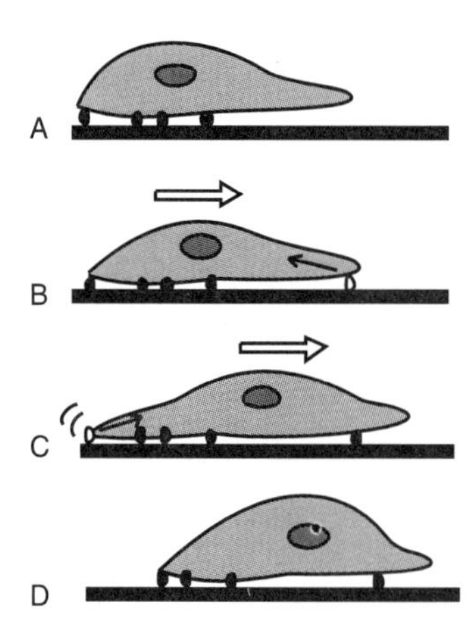

Figure 5.6. Schematic of the process of cell migration. In **A**, the cell extends a protruding lamellipodium. In **B**, the lamellipodium has formed an adhesive contact on the substrate surface; the cell generates traction by pulling against that adhesive contact to move the bulk of the cell forward. In **C**, the cell has moved forward such that cell-substratum adhesive contacts at the rear of the cell are now being pulled against as the cell continues to move. In **D**, the rear cell-substrate adhesion has been broken, and the cell is ready to project more lamellipodia and continue the process. For some very fast-moving cells (such as macrophages), if the adhesive contacts at the rear of the cells are not dissolved quickly enough to keep pace with the rapid forward movement, the cell will actually "rip away" from these adhesive contacts, leaving behind small focal deposits of membrane and adhesion complexes as the cell migrates forward.

can be a stimulus for the cell to migrate in that general direction. The process of cell migration starts with the extension of lamellipodia in the direction the cell will move (Fig. 5.6). The lamellipodia are formed by polarized polymerization of the actin filaments that make up the cell's cytoskeleton, essentially lengthening the cytoskeleton in a particular direction. The cell then makes tight adhesive contacts with the substratum underneath that protruding portion of the cell, which provide stable anchors for the newly polymerized cytoskeletal filaments. The cell can then pull against the substrate, moving the cell body forward. The actin filaments at the rear of the cell depolymerize, and cell-substrate adhesive bonds at the rear of the cell are broken, allowing overall cellular motion forward.

It's important to note that even though this description of chemotaxis and migration is given in reference to tissue macrophages, the general mechanisms of cell migration given here apply to many different kinds of cells including the neutrophils—the body's second line of defense in the inflammation process—which also migrate to injury sites.

What types of chemicals stimulate the chemotactic migration of macrophages and neutrophils? There is a wide variety of *chemoattractants* for these cells, including but not limited to chemicals released from injured tissues, like *tissue plasminogen activator*; chemicals that are products of the coagulation process, like *kallikrein* and *prostaglandins*; chemicals from fibrinolysis like *fibrin degradation products*; and members of the *lymphokine* class of chemicals, which are

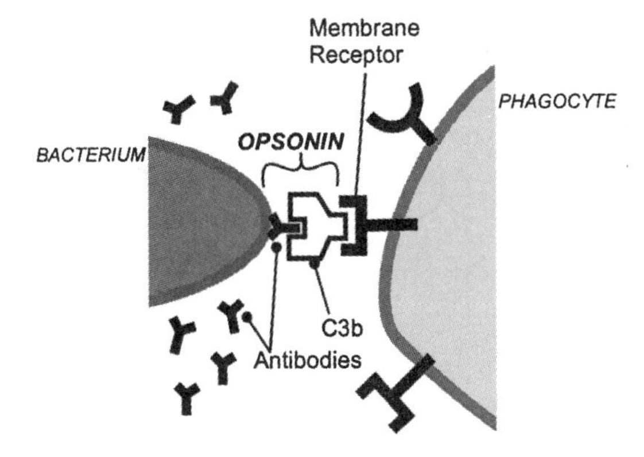

Figure 5.7. Schematic of opsonization. Antibodies from the body's immune system attach to foreign objects (in this example, a bacterium); a chemical called C3b then attaches to the immobilized antibody, and the antibody-C3b complex is termed opsonin. Opsonin interacts with receptors on the cell membrane of a phagocyte (macrophage or neutrophil), stimulating the process of phagocytosis.

produced by cells from the immune system and are discussed in Chapter 6. Because these chemicals are chemoattractants for leukocytes and are produced from injury or coagulation, they provide a direct link between the first two phases of the wound healing process (coagulation and inflammation) and help explain why inflammation is a natural, necessary part of the wound healing process.

5.4.2 Phagocytosis

Once the tissue macrophages encounter foreign objects (particles of dirt or debris, microorganisms, pieces of damaged tissue or dead cells) at the injury site, these objects are phagocytosed. What exactly does that mean? How does a macrophage or a neutrophil recognize that an object is foreign and needs to be destroyed?

Most normal, healthy structures in the body have a smooth surface. Roughness of an object indicates that it is out of place and increases the probability of consumption. Additionally, the body's immune system chemically "marks" foreign objects (this process is discussed in detail in Chapter 6) with antibodies for phagocytosis. For example, if bacteria enter a wound, antibodies will attach to the bacterial membrane; a chemical called C3b then binds both to the antibody and to receptors on the neutrophil or macrophage membrane, initiating phagocytosis (Fig. 5.7). This process is called *opsonization*.

Once the neutrophil or macrophage recognizes and attaches to an opsonized object, the cell projects pseudopodia around the object. These pseudopodia meet and fuse, engulfing the object in a vacuole called a phagosome (Fig. 5.8). Proton pumps in the vacuole membrane reduce the pH in the phagosome to approximately 4.0. This acidic environment can quickly kill some microorganisms, and it enhances the activity of some lysosomal enzymes. Meanwhile, granules in the

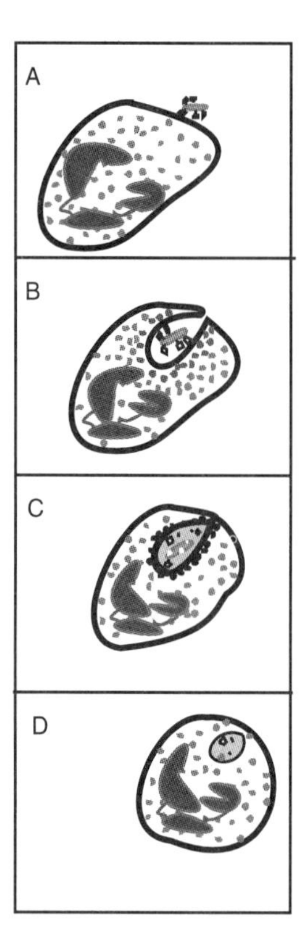

Figure 5.8. The process of phagocytosis. In **A**, a neutrophil interacts with an opsonized bacterium, and activation of the neutrophil membrane receptors initiates the phagocytic response. In **B**, the neutrophil is extended around the bacterium, forming a membrane-encapsulated phagosome. Granules in the neutrophil are starting to move toward the phagosome. In **C**, the granules are discharging into the phagosome during the metabolic burst, and in **D**, remaining pieces of the bacterium are digested.

leukocyte converge on and discharge their contents into the phagosome. Enzymes from the granules degrade and digest lipids and proteins; the granules also release bactericidal agents (like major basic protein and lactoferrin, which binds the iron necessary for bacterial nutrition). At the same time, the cell experiences a "metabolic burst," in which glucose metabolism increases up to tenfold and oxygen consumption increases by two- or threefold. The metabolic burst allows the formation of superoxide anions and hydrogen peroxide in the phagosomes. These reactive radicals and strong oxidizers can inactivate proteins as well as kill bacteria, viruses, yeast, and fungi. Remaining material in the phagosome is digested, and the neutrophil or macrophage moves on to the next foreign object. A neutrophil can usually phagocytose between 5 and 20 bacteria before it is essentially "worn out" and dies, whereas a macrophage can phagocytose much more than a neutrophil—up to 100 bacteria—before it dies. Macrophages can also phagocytose much larger particles (up to the size of a red blood cell) than neutrophils are able to handle. Figures 5.9 and 5.10 show macrophages in the process of phagocytosing particles of high-density polyethylene.

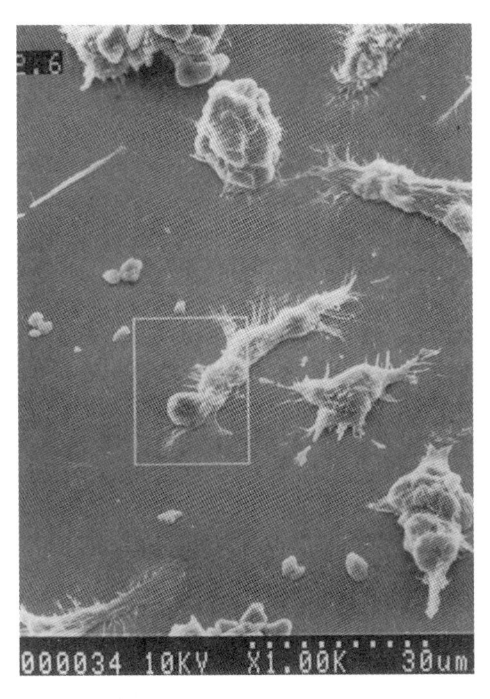

Figure 5.9. Scanning electron micrograph of macrophages phagocytosing polymeric particles. In the center of the figure, the box designates a macrophage just beginning to phagocytose a large particle of high-density polyethylene. In the center top and lower right of the figure are macrophages that have already phagocytosed a number of particles. Figure reproduced from I. Voronov, J.P. Santerre, A. Hinek, J.W. Callahan, J. Sandhu, E.L. Boynton, "Macrophage phagocytosis of polyethylene particulate in vitro," *Journal of Biomedical Materials Research* 39:40–51 (1998), © John Wiley & Sons, Inc., 1998. Reprinted by permission of Wiley-Liss, Inc., a subsidiary of John Wiley & Sons, Inc.

One relevant question to ask at this point is, What happens if the particle ingested by a macrophage is essentially indigestible, like an asbestos fiber or a silica particle? If the phagosome is unable to break down the ingested object, intracellular lysosomes fuse with the phagosome and eventually digest thc phagosome itself. The toxic contents of the phagosome spill into, and kill, the macrophage. Lysis of the macrophage releases not only the original indigestible object for reuptake by other macrophages, but also the chemical contents of the macrophage. If the number of indigestible objects is relatively small, the process of particle uptake, macrophage death, and reuptake of particles can continue for long periods of time, with slow waves of macrophages continually lysing and being replaced. If the number of indigestible particles is large, and a large number of macrophages are lysing and being replaced by new cells (which will also lyse), this continual stimulus can escalate to cause clinical problems, as is the case in *silicosis*. Some of the chemical products released from dying macrophages are stimulants for the proliferation and function of *fibroblast* cells, which essentially produce collagenous soft tissue in the body. Silicosis is a disease in which silica particles inhaled into the lungs cause large amounts of macrophage

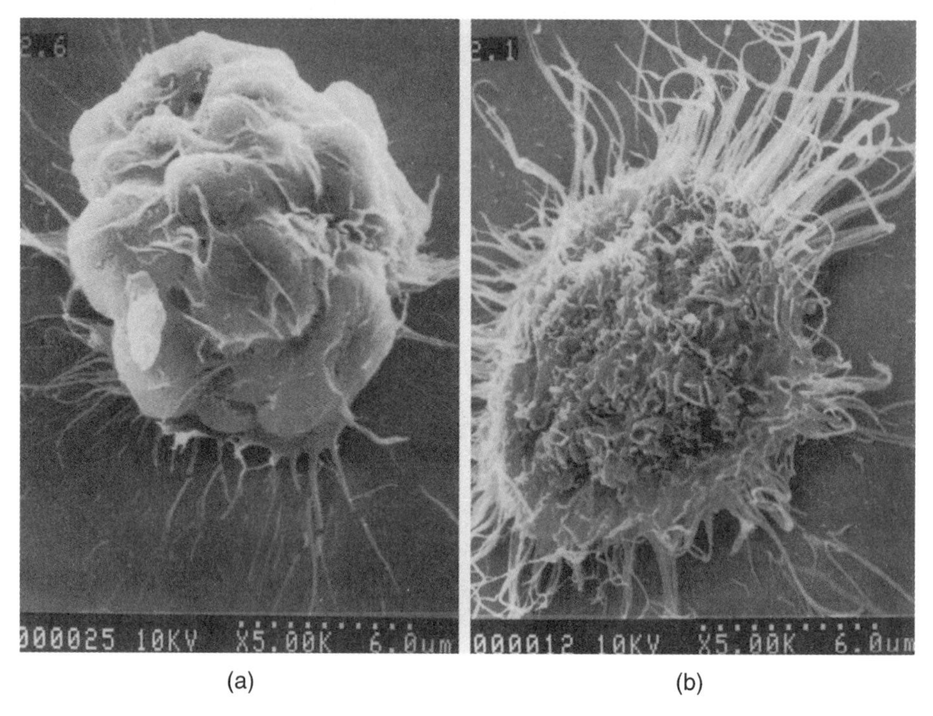

Figure 5.10. Scanning electron micrographs of macrophages that have and that have not ingested particulate debris. The shapes of the ingested high-density polyethylene particles have distended the macrophage in the left panel to such a degree that the cell resembles one giant globular particle. The macrophage in the right panel was not exposed to particulate debris, and it exhibits a typical, normal cellular shape and surface morphology. Figure reproduced from I. Voronov, J.P. Santerre, A. Hinek, J.W. Callahan, J. Sandhu, E.L. Boynton, "Macrophage phagocytosis of polyethylene particulate in vitro," *Journal of Biomedical Materials Research* 39:40–51 (1998), © John Wiley & Sons, Inc., 1998. Reprinted by permission of Wiley-Liss, Inc., a subsidiary of John Wiley & Sons, Inc.

lysis and subsequent fibroblastic stimulation; the resulting fibrotic regions in the lung can severely impair oxygenation.

5.4.3 Diapedesis

Immediately after an injury the processes of coagulation and inflammation begin, and the tissue macrophages ("the first line of defense") located near the injury quickly move to the site of injury and begun phagocytosing any foreign objects and damaged cells/tissues. The body's second line of defense after an injury consists of the neutrophils circulating in the bloodstream. Some of the chemical products of injured tissues interact with the endothelial cells lining the blood vessels near the injury. This causes the endothelial cells to express more cell-cell adhesion receptors on their surface and also allows the endothelial cells to retract slightly from each other. The increase in adhesion receptors on the endothelium is matched by an increase in adhesion receptors on circulating neu-

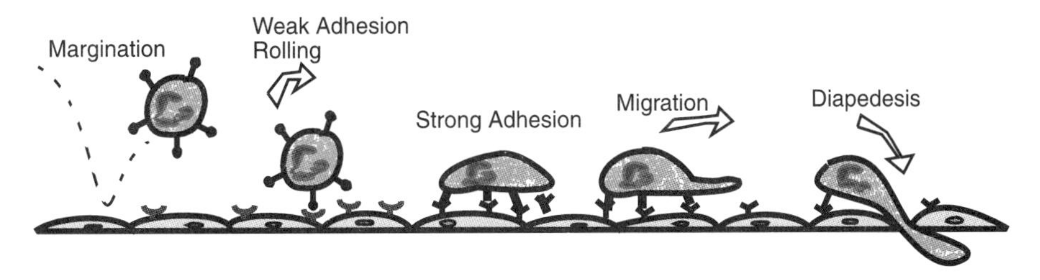

Figure 5.11. Schematic of a leukocyte moving from the bloodstream into tissues. The cell-surface receptors are enlarged in this figure to show and emphasize the importance of cell-cell adhesive contacts to this process. At the left of the figure is the first part of the process, margination, in which a neutrophil moves to the side of the blood vessel wall, very weakly adhering and then "bouncing away" until the neutrophil-endothelium adhesion is strong enough to capture the neutrophil from the bloodstream but still weak and/or dynamic enough to permit rolling of the neutrophil along the endothelium. The adhesion receptors that dominate these weak and/or dynamic adhesive events are the *selectins*. When the neutrophil contacts the endothelium, the selectin receptors are shed from cell membrane and stronger adhesion receptors, *integrins*, are expressed instead. The integrins allow the neutrophil to adhere strongly to the endothelium, stopping the rolling process and allowing the neutrophil to migrate over the surface of the endothelium. The neutrophil can then squeeze between endothelial cells (diapedesis), during which receptor-ligand interactions between the neutrophil and endothelial cells "loosen" the otherwise tight endothelial junction, causing the endothelial cells to retract slightly. This allows the neutrophil to migrate out of the blood vessel and into the tissues with minimal coagulation initiation from blood-collagen contact.

trophils. Instead of bumping into the endothelium and then bouncing back off into the bloodstream, the neutrophils start to stick lightly to the endothelium and roll along the endothelium (Fig. 5.11). This is called *margination*, because the neutrophils are essentially being relocated to the margin of the bloodstream. Near the site of injury, the kinds and amounts of adhesion receptors expressed on the endothelium (and the neutrophils) change to encourage strong, rather than weak, adhesion of neutrophils. The neutrophils adhere to the endothelial surface, migrate over the endothelium to find a junction between two endothelial cells that is somewhat flexible, extend a pseudopod between the retracted endothelial cells, and proceed to squeeze themselves through the endothelial layer. This tight squeeze through the endothelial junction and into the underlying tissue is called *diapedesis*. The neutrophils then migrate through the tissue to the site of injury and proceed to phagocytose particles of foreign matter or damaged cells/tissues.

Chemical products of the inflammation process that enter the bloodstream and reach the bone marrow stimulate the release of more neutrophils into the blood. After an injury, the circulating level of neutrophils can rise as much as four- to fivefold, making more neutrophils available to combat the injury.

The body's third line of defense in the inflammation process is the monocyte-macrophage system. Although some tissue macrophages near the injury site can be mobilized quickly, the total number of locally available tissue macrophages is generally low. Recall that monocytes circulate in the blood in an immature form. Monocytes move from the blood vessels into the tissues in the same manner as neutrophils, and as the monocytes migrate through the tissue they mature into full-fledged macrophages. It takes at least 8 h for migrating monocytes to achieve full phagocytic capacity, so the neutrophils reach the injury site much more quickly and are effective before the blood-derived macrophages can mature and function optimally. Moreover, the number of monocytes circulating in the blood and stored in the marrow is lower than the number of available neutrophils, putting the monocyte-macrophage system at the third level of the body's defense system. The fourth and final level of the defense system is the process of upregulating production of monocytes and granulocytes in the bone marrow. The marrow is capable of producing large quantities of these cells on demand, but it takes 3–4 days from the initial marrow stimulus to the time that new granulocytes and macrophages enter the bloodstream. Therefore, the rapid tissue macrophage and neutrophil response, as well as the mid-stage monocyte-macrophage response, are crucial to ensuring that any microorganisms in the wound do not have a chance to establish a thriving colony.

5.5 PHYSIOLOGICAL EXPLANATIONS FOR THE CARDINAL SIGNS

We can now combine our knowledge of coagulation with our understanding of leukocyte function to explain why some degree of inflammation is invariably a consequence of coagulation and what physiological process give rise to the "cardinal signs" presented in Section 5.2. The first stages of inflammation are characterized by local vasodilation and increased vascular permeability. Many of the chemicals that initiate these processes are derived from coagulation and are mentioned in Chapters 4 and 5, for example, bradykinin, serotonin, and prostaglandins. Other mediators of these processes are discussed in Chapter 6, notably, parts of the *complement system* and *lymphokines*. Kinins in particular are strong vasodilators, so the surface-mediated activation of factor XII (leading to the production of kallikrein and its subsequent role in kinin formation; Fig. 4.29) is an initial trigger for the onset of inflammation. The vasodilation brings more blood into the capillaries near the injury site, while at the same time the endothelium becomes more adhesive, trapping circulating leukocytes and slowing blood flow locally. The increased number of slow-moving erythrocytes in the area gives a reddish color to the tissue, and the added blood volume in the area can make the tissue feel warm. More warmth can be generated locally by the large numbers of leukocytes infiltrating the area and operating at increased metabolic rates (recall the "metabolic burst" in phagocytosis) as well as by the presence of fever-causing chemicals called *pyrogens*. Pyrogens can be released from damaged tissues or secreted by bacteria; moreover, when leukocytes phagocytose bacteria or bacterial products (especially *endotoxins*), the leukocytes release a chemical called *interleukin-1*, which is a pyrogen.

TABLE 5.2. Physiological Explanations for the Cardinal Signs of Inflammation

Cardinal Sign	Physiological Causes
Rubor	• Vasodilation, adhesive endothelium, clotting locally concentrates erythrocytes
Tumor	• Vasodilation and increased vascular permeability (from bradykinin, prostaglandins, histamine, etc.) cause local influx of fluid and cells
Calore	• Vasodilation and increased vascular permeability increase local blood volume • Large numbers of leukocytes operating at increased metabolic rates • Pyrogens
Dolore	• Deep pain receptors activated by local swelling (throbbing pain) • Sharp pain from injury, kinins

This table summarizes links between the processes of coagulation and inflammation, without considering contributions from the immune system and the complement cascade (discussed in Chapter 6).

Coagulation-generated kinins play another role in inflammation by increasing vascular permeability (histamine from leukocytes also has this effect), essentially "loosening" the adhesion of endothelial cells to one another. This is important because it makes the process of leukocyte diapedesis feasible. The increased vascular permeability also allows the flow of water and blood proteins from the circulation into the tissues. This fluid influx causes local tissue swelling, especially if the local lymphatic system is blocked or constricted because of the initial injury. Interstitial blood clotting may occur once blood proteins move past the vessel wall and encounter connective tissue. This slows the drainage of fluid from the area and can also delay the spread of any infectious organisms from the wound site into the tissues.

During these chemically initiated inflammatory events, the body's cellular defenses are mobilized and the tissue macrophages, neutrophils, and monocyte-macrophages move to the injury site to perform their functions. The cardinal signs of inflammation can now be listed with corresponding physiological explanations for each sign (Table 5.2). At this point, we have a fairly complete picture of why inflammation is a normal, necessary part of wound healing that will assuredly arise, to some degree, from the process of coagulation. We are only missing chemical and cellular contributions from the immune system, which are the focus of Chapter 6.

5.6 INFECTION

If bacteria, fungi, or viruses manage to colonize a wound, the wound is *infected.* The beginning of this chapter made a distinction between inflammation and infection, noting that inflammation generally accompanies infection, but infection does not have to accompany inflammation. One distinguishing feature of infection is the production of pus, or suppuration. If bacteria (as an example) enter the tissue in great enough numbers, the short-term inflammatory response may be unable to eradicate the bacterial colony quickly. Massive numbers of neutrophils and macrophages at the infection site will phagocytose as many bacteria as possible and then die, producing a liquefying mass of dead cell products, bacteria, and *necrotic* (dead) tissue: This liquid is pus. If the next and larger wave of monocyte-macrophages and neutrophils is able to successfully destroy the bacteria, the accumulated pus may be resorbed and phagocytosed over a time period of days, leaving little permanent tissue damage. If the infection is sustained over time, fibrous tissue may form around the liquefied infectious region, to encapsulate the bacteria and prevent their spread to other parts of the body. This *abscess* often works its way to the surface of the tissue, breaks open, and drains, collapsing the abscess and removing most of the source of infection. Why do abscesses move to the tissue surface? As abscesses grow (either from expanding liquefaction or from more fibrous tissue encapsulation) they grow in the direction that offers the least resistance to movement, which (for surface wounds) is generally toward softer, fatty tissues and the skin. On reaching the skin, the abscess stretches the skin until it bursts open or is surgically drained. If the abscess has no clear path of least resistance to follow (for example, it is deep within the body or at a joint), it may grow and do a good deal of local damage in its site. Bacteria from a trapped abscess may even enter the bloodstream and colonize a new site, spreading the infection to entirely different parts of the body. Infections are a serious concern when considering the surface properties of biomaterials and implantation procedures.

In his book *Biological Performance of Materials*, Jonathan Black discussed three types of infection associated with implants [3], which we will briefly present here. First, *superficial immediate infections* occur on or near the skin and can be caused by bacteria that normally populate the skin or that are airborne. Infected sutures are an example of the superficial immediate type. The second type of infection is the *deep immediate infection*. This infection type tends to occur shortly after invasive surgeries, such as implantation procedures, and seems to be primarily due to the inadvertent relocation of skin bacteria into the body. Airborne bacteria may also play a role in this type of infection. The third type of infection, the *deep late infection*, occurs months or years after surgery in areas that had previously showed no signs of trouble. It is thought that two mechanisms may be responsible for these infections. First, the infection may simply be a delayed display of contamination that actually occurred during surgery but was slow to develop. Second, the infection may be the result of bacteria that were present at another anatomic site (for example, an infected cut), which entered the bloodstream and were transported to and colonized a new site. Deep late infections are a serious concern in many implantation procedures, not because of the frequency of occurrence (which is low) but because of the difficulty in treating these infections [3]. Infected sutures or burn dressings can easily be removed and replaced with new, clean materials. Removing and replacing a heart valve or hip implant is much more difficult. Local application of topical antibiotics can be used to treat superficial infections, and mild systemic antibiotics may be used to treat deep immediate infections because the microorganisms have not had the time to develop large colonies. Treating a deep late infection with strong systemic antibiotics can have unpleasant side effects and may not solve the underlying problem.

Many researchers are currently working to design biomaterials and implanted devices that discourage bacterial adhesion and growth. For example, geometric factors in implants and implantation sites can be modified, avoiding the production of "dead space" (space free of tissue, which typically fills with fluid in vivo) that provides a sheltered location for bacterial colonies. Porous materials, which by their nature include small "dead spaces," have been examined with respect to infection potential, but the combined data are somewhat inconclusive because the chemistry (and corrosion products) of different biomaterials can also affect bacterial colonization. Biomaterials designed to release antibiotics via diffusion or dissolution of the material may provide some protection against deep infections, and these types of materials have been used clinically. However, there are some potential difficulties with this approach [3]. Without knowing the type(s) and activity level of the bacteria that may infect an implantation site, selecting an appropriate kind and amount of antibiotic is difficult. The antibiotic may not diffuse far enough from the implant to fully protect all of the tissue at risk of infection. If the patient developed a sensitivity to the antibiotic, the implanted biomaterial would need to be removed, and that is not always possible in critical implant applications. Furthermore, the widespread use of antibiotics has led to the emergence of antibiotic-resistant bacterial strains. These drug-resistant bacteria would not be affected by a biomaterial that released a typical antibiotic. Additionally, exposing individuals to antibiotics that are not absolutely necessary, or to lower-than-effective levels of antibiotics, increases the risk of generating additional drug-resistant strains of bacteria.

5.7 SUMMARY

- Historically, inflammation and infection often occurred simultaneously in wounds. However, these are two separate processes, with distinct causes, and should be differentiated from each other.
- Physiological causes of the four cardinal signs of inflammation, "*rubor et tumor cum calore et dolore*," can be explained on the cellular and molecular levels, helping to show why some degree of inflammation is invariably a consequence of blood-biomaterial contact and subsequent coagulation.
- The body's first line of defense against pathogens is the tissue macrophages; circulating neutrophils are the next to arrive at a wound; and the body's third line of defense is the macrophages produced from circulating monocytes. The long-term presence of foreign materials in the body may result in increased production of monocytes and granulocytes in the bone marrow and the formation and continued presence of foreign body giant cells at the tissue-material interface.

- Understanding the molecular level processes involved in cell migration, phagocytosis, and diapedesis helps to link the coagulation and inflammation stages of wound healing as well as to link functions of nonlymphatic and lymphatic leukocytes (discussed in Chapter 6).
- Because of the problems associated with treating implant-associated infections, many researchers are currently working to design biomaterials and implanted devices that discourage bacterial adhesion and growth. Although biomaterials that release antibiotics could help reduce the number and severity of implant-associated infections, the injudicious use of antibiotics could possibly create other clinical problems.

5.8 REFERENCES

[1] G.B. Ryan and G. Majno, *Inflammation*, published by the Upjohn Company, Kalamazoo, MI (1980).

[2] R.S. Cotran, T. Collins, V. Kumar, *Robbins Pathologic Basis of Disease* (6th edition), W.B. Saunders Company, pg. 52 (1999).

[3] J. Black, *Biological Performance of Materials Fundamentals of Biocompatibility* (3rd edition revised and expanded), Marcel Dekker, Inc., pgs. 137–138 (1999).

5.9 BIBLIOGRAPHY/SUGGESTED READING

Black, J., "The Inflammatory Process," in *Biological Performance of Materials Fundamentals of Biocompatibility*, third edition revised and expanded, Marcel Dekker, Inc., New York, NY (1999) pp. 131–153.

Guyton, A.C., "Chapter 33: Resistance of the Body to Infection: I. Leukocytes, Granulocytes, the Monocyte-Macrophage System, and Inflammation," in *Textbook of Medical Physiology*, eighth edition, W.B. Saunders Company, Harcourt Brace Jovanovich, Inc., Philadelphia, PA (1991), pp. 365–373.

Hayhoe, F.G.J. and Flemans, R.J., *Color Atlas of Hematological Cytology*, third edition, Mosby-Year Book, Inc., St. Louis, MO (1992).

Junquiera, L.C., Carneiro, J., and Kelley, R.O., "Chapter 12: Blood Cells," in *Basic Histology*, eighth edition, Appleton & Lange, Norwalk, CT (1995), pp. 222–233.

Lauffenburger, D.A., and Linderman, J.J., *Receptors: Models for Binding, Trafficking, and Signaling*, Oxford University Press, New York, NY (1993).

5.10 STUDY QUESTIONS

1. Explain the difference between inflammation and infection, and describe why infection and inflammation were often historically confused with each other.

2. Describe the progression of events as an infected wound heals and as an infected wound forms an abscess.

3. Recite and translate the "cardinal signs" of inflammation.

4. List the types of terminally differentiated nonlymphatic leukocytes and describe their main functions, including (where appropriate) important, physiologically active chemicals secreted.

5. Explain how monocytes, macrophages, and foreign body giant cells are distinctly different and yet related.

6. Describe the differences between macrophage and neutrophil functions.

7. Sketch and explain how cell migration is dependent on the dynamic functions of adhesion receptors.

8. Define chemotaxis and provide examples of leukocyte chemoattractants that are important to the processes of coagulation and fibrinolysis.

9. Describe opsonization and the process of phagocytosis when the particles to be phagocytosed are digestible and when the particles are indigestible.

10. Describe the roles of tissue macrophages, circulating neutrophils (including the process of diapedesis), and the monocyte-macrophage system in discouraging the establishment of an infection in a wound.

11. Give physiological causes for the cardinal signs of inflammation, based on an understanding of the cellular and molecular components of coagulation and inflammation.

5.11 DISCOVERY ACTIVITIES

1. If you were a doctor and were given a set of clinical symptoms from patients with varying kinds of infections, what questions would you ask or what would you look for to discriminate between superficial immediate, deep immediate, and deep late infections?

2. There are a number of excellent collections of color images of the cells discussed in this chapter. Printed atlases of hematology, cytology, and pathology can be found at medical school libraries, and a number of medical schools and professors have made good collections of histology/pathology images publicly accessible on the World Wide Web (often with accompanying short explanations or tutorials). Find a printed atlas or a website, look at color images of the hematopoietic cells, and compare them to the black-and-white sketches of these cells presented in this textbook. What main features do the sketches try to convey? What features are missing? Explore how the shapes of hematopoietic cells change with various diseases.

3. Knowing what you know now about the cellular and molecular basis of inflammation, if you were to badly sprain your ankle, would you apply ice to the injury or heat? Why?

4. Take a look at published proceedings from a recent scientific conference of a professional society that focuses on materials science and/or biomaterials (the Society For Biomaterials, for instance, or the Materials Research Society). Find three abstracts or papers that present recent research on modification of the surfaces of materials to discourage bacterial adhesion. Are there common surface modification strategies or principles among the papers/abstracts you've selected? Do the results of the three studies agree with or complement each other? If you were the author of all three studies, what sort of experiments would you perform next?

5. Do a literature search to find two related journal articles about mathematical models of cell migration. What types of physical parameters (number of cells, strength of adhesion receptor-ligand interaction, etc.) are represented in the models? Which parameters seem to have the largest effect on the overall behavior of the model? What assumptions are inherent in the models, and how do they compare to conditions that might be encountered in the body?

6

The Immune System and Inflammation

6.1 INTRODUCTION

The immune system is a remarkable and complex organization of defensive strategies. Immunology is a broad field of study, with many specialized subfields of research and development. This textbook focuses only on fundamental information about the immune system, specifically, links between the immune system and the inflammation process. Even within this small focus area, many mechanisms of the body's immune response are topics of current study and debate. There are a number of excellent texts devoted to the subject of immunology (a few of which are listed in Section 6.8, Bibliography/Suggested Reading), and we encourage students to explore this fascinating field of research beyond the information presented here.

The human body has two general kinds of immunity, or resistance to foreign organisms. The first, *innate immunity*, results from a range of general processes that generally make it difficult for pathogens to enter the body and establish a stronghold. The skin's function as a protective barrier, the acidic and enzymatic environment of the stomach, enzymes in the blood and other body fluids that lyse or inactivate bacteria, and even neutrophil/macrophage phagocytosis of foreign objects are all general defensive mechanisms considered part of innate immunity. The term *acquired immunity* is used to describe mechanisms of resistance that are developed within the body in response to specific, individual foreign organisms or stimuli. Acquired and innate immunity combine to produce a defensive system that has the following characteristics:

1. Specificity—The system is able to discriminate between and respond to a large number of stimuli;
2. Memory—After an initial exposure to a stimulus, subsequent encounters with that stimulus provoke a stronger and faster response than that which occurred during the first exposure;
3. "Self-knowledge"—Foreign invaders, including biomolecules, cells, and tissues from other species (or sometimes from other individuals within the

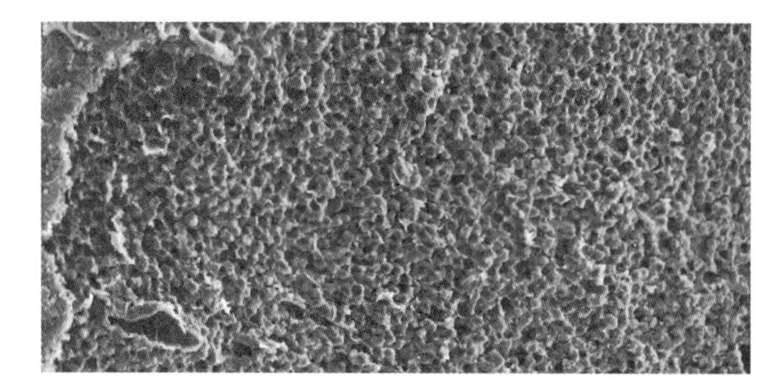

Figure 6.1. Scanning electron micrograph of the interior of a rat thymus. This picture shows a fairly uniform structure of lymphocytes within a framework of epithelial-reticular cells. Original magnification: ×230. Figure reproduced from page 506 of L. Weiss, *The Blood Cells and Hematopoietic Tissues* (1977), © McGraw-Hill, Inc., 1977. Reprinted with the permission of The McGraw-Hill Companies.

same species), are rapidly recognized and attacked, whereas biomolecules, cells, and tissues of the host are ignored.

There are two subtypes of acquired immunity: *humoral immunity*, which arises from the chemical activities of antibodies, and *cell-mediated immunity*, which arises from the cytotoxic actions of cells. Both of these subtypes of immunity result from actions of the two main classes of lymphocytes.

6.2 LYMPHOCYTES

The exact sequence of lymphocyte precursor cell development and differentiation is still unclear. It is known that lymphoid stem cells in the bone marrow produce daughter cells that commit to separate pathways for *B lymphocytes* or *B cells*, which develop in the bone marrow until mature, and *T lymphocytes* or *T cells*, which are developed in the thymus (Figs. 6.1 and 6.2) from precursor cells that leave the marrow and arrive in the thymus via the bloodstream. Both B cells and T cells are released into the bloodstream when mature and migrate to nonthymic lymphoid tissues (such as the spleen, tonsils, or lymph nodes). A third type of lymphocyte, *natural killer cells* or NK cells, appears also to be derived from a bone marrow precursor, but little beyond that is known about NK cell production. The exact physiological roles of NK cells are still unclear, but it is thought that these cells are able to recognize and destroy tumor cells and possibly cells infected with certain kinds of viruses. NK cells are sometimes considered to be one of the innate immunity mechanisms rather than the acquired immunity mechanisms, because it is unclear how or whether these cells develop responses to specific stimuli. In an adult and under normal conditions, NK cells make up approximately 10–15% of the lymphocytes circulating in the

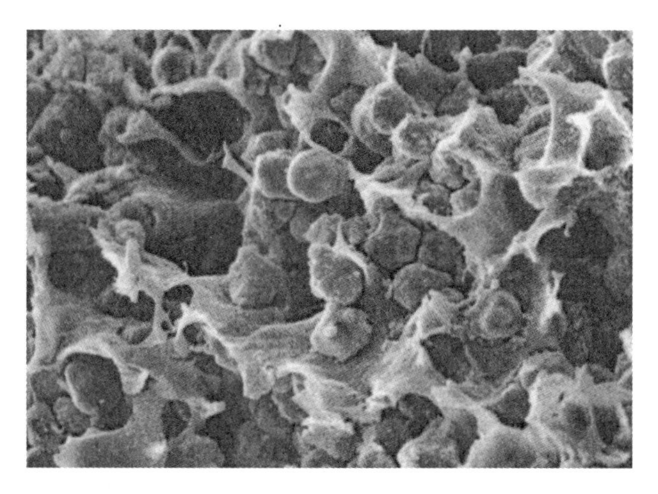

Figure 6.2. High-magnification scanning electron micrograph of rat thymus tissue. Many of the round lymphocytes have been removed to expose the underlying structure of the epithelial-reticular cells. Figure reproduced from page 507 of L. Weiss, *The Blood Cells and Hematopoietic Tissues* (1977), © McGraw-Hill, Inc., 1977. Reprinted with the permission of The McGraw-Hill Companies.

bloodstream, whereas B cells represent between 5% and 10% and T cells represent 65–75% of the circulating lymphocytes. B cells are responsible for humoral immunity, whereas T cells are considered to be responsible for cell-mediated (acquired) immunity. To explain the functions of B and T cells clearly, some new terminology must first be introduced and clarified.

6.3 IMMUNOGENS, ANTIGENS, AND ANTIBODIES

The term pathogen has been used previously in this textbook to generally describe a disease-causing agent. An *immunogen* is a molecule or collection of molecules that can induce a response from the body's immune system. Bacteria, viruses, pollen from grass or trees, and proteins in food are all examples of possible immunogens. So, although there is a good chance that a given pathogen will probably act as an immunogen, an immunogen is not necessarily always a pathogen. Table 6.1 summarizes some general traits of immunogenic molecules.

The term *antigen* is used to refer to substances that are specifically recognized by an *immunoglobulin* (a specific kind of glycoprotein, explained below) or by receptors on the T cell membrane. Many, but not all, antigens trigger an immune response. Therefore, whereas all immunogens are also antigens, it is possible for some compounds to be recognized by the immune system (antigenic) but not provoke an immune response (nonimmunogenic), so an antigen is not always an immunogen. The specific chemical site on an antigen that binds to

TABLE 6.1. Factors important to determining degree of immunogenicity.

Immunogenic Factor	Characteristics
Chemical Composition	• Large macromolecular proteins tend to be potent immunogens • Polysaccharides, short polypeptides, and some organic polymers can be immunogens
Molecular Size	• Smaller molecules (molecular weight less than 10,000) tend to be weakly or non-immunogenic • Large proteins (molecular weight above 100,000) are the most potent immunogens • Exceptions to these cutoff molecular weights exist
Chemical Complexity	• In general, immunogenicity increases with structural and chemical complexity
Genetics of Host Animal	• Different species are genetically conditioned to respond in various ways to potential immunogens (example: pure polysaccharides are immunogenic in mice and humans but not rabbits) • Immune tendencies can be inherited
Degree of Foreignness	• Some biomolecules will not provoke an immune response when transplanted between members of the same species, but will evoke an immune response when transplanted between members of different species
Method of Administration	• A substance that is not immunogenic when injected intravenously may be immunogenic if injected subcutaneously • The threshold dose required to elicit an immune response varies from compound to compound; larger doses do not necessarily translate to larger immune responses

Information summarized from Goodman, J.W., in *Basic & Clinical Immunology*, eighth edition (Stites, D.P., Terr, A.I. and Parslow, T.G., eds.), Appleton & Lange, Norwalk, CT (1994), pgs. 50–51.

an immunoglobulin or T cell receptor is called an *epitope*, and sometimes also called an *antigenic determinant*. A single antigen may have multiple epitopes, yielding multiple sites for immunoglobulin or T cell receptor binding (Fig. 6.3).

Immunoglobulins are a class of glycoproteins that bind to specific epitopes on antigens. There are five classes of immunoglobulins in humans (Table 6.2), often called *antibodies*. Figure 6.4 shows the general structure of the immunoglobulin G (IgG) molecule; most of the antibodies found in human blood are of the IgG type. There are variations in structure between the five Ig classes, but all classes have pairs of heavy and light chains, and all heavy-light chain pairs have variable and constant portions. The variable portions of antibodies possess a remarkable variety of conformations and amino acid compositions, allowing a wide range of antibodies to exist within a given Ig class and allowing a given antibody to selectively bind to a specific target antigen. Antibody-antigen binding is very strong and sterically specific, relying on the combined actions of hydrophobic bonding, hydrogen bonding, ionic attractions, and van der Waals forces.

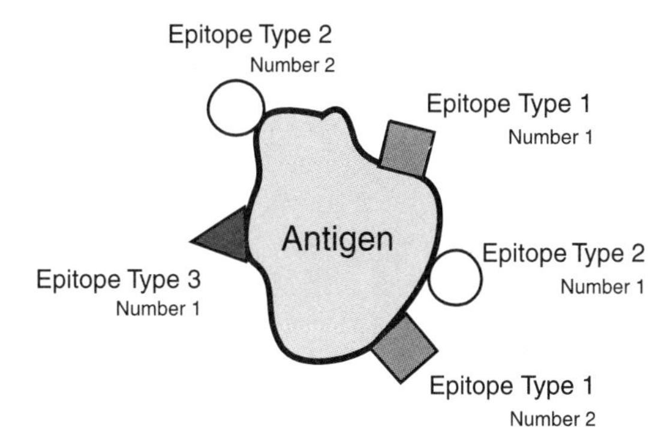

Figure 6.3. Schematic of a single antigen with multiple epitopes. This hypothetical cross-section through some form of antigen is intended to demonstrate that one antigen may possess multiple type of epitopes, allowing multiple types of immunoglobulins or T cell receptors to bind to the antigen. A single antigen may also possess copies of the same epitope, allowing a number of the same type of immunoglobulins or T cell receptors to bind to the antigen.

The immune system's response to relatively weak immunogens can be enhanced if the immunogen is mixed with *adjuvants* or *haptens. Adjuvants* have their own immunogenic properties, resulting in a stronger local stimulus to the immune system. Adjuvants may be chemicals or may simply be particles that adsorb the weak immunogen, increasing the effective size of the weak immunogen (thus enhancing phagocytosis of the particle by macrophages) or pro-

TABLE 6.2. Characteristics of the immunoglobulin classes.

Immunoglobulin	Characteristics
IgG	• Constitutes 75% of plasma immunoglobulins • The only immunoglobulin that crosses the placenta to circulate through the fetal bloodstream
IgA	• Small amounts in serum, but the main immunoglobulin in tears, colostrum, saliva, as well as nasal and bronchial secretions • In general, secreted in mucosal membranes to provide protection against the proliferation of microorganisms in bodily secretions
IgM	• Constitutes 10% of serum immunoglobulins • One of two major immunoglobulins expressed on B lymphocytes • Effective activator of the complement system
IgE	• High affinity for membrane receptors of mast cells and basophils • Antigen-antibody complexes formed on the surface of mast cells and basophils bioactive substances such as histamine and heparin • "Allergic reactions" are mediated by IgE-antigen interactions
IgD	• One of two major immunoglobulins expressed on B lymphocytes • Properties and functions not well understood, appears to be involved in B lymphocyte differentiation

Information summarized from Junquiera, L.C., Carneiro, J., and Kelley, R.O., *Basic Histology*, eighth edition, Appleton & Lange, Norwalk, CT (1995), pgs. 247–249.

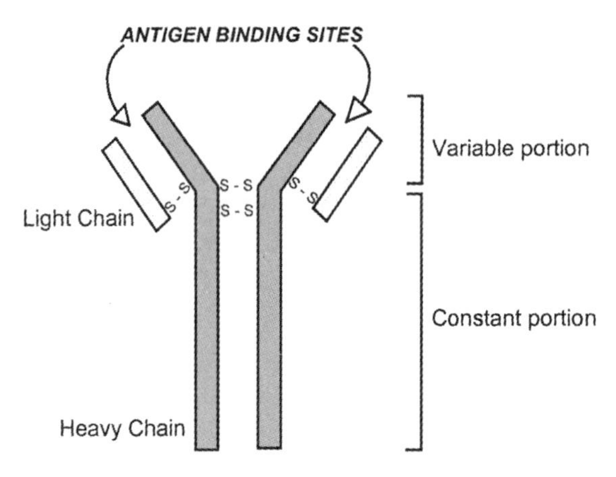

Figure 6.4. Schematic of the structure of an IgG molecule. The IgG structure is often given as a model for all the immunoglobulins. Two long "heavy chains" form the inside of a "Y"-shaped structure, with two short "light chains" filling in the arms of the "Y." Sulfur-sulfur bonds hold the chains together. Variations in the composition (and thus ionic and steric structure) of the "variable portion" of the molecule allow epitope-specific binding.

longing local retention of the weak immunogen (thus increasing the chances of a local immune response). *Haptens* are low-molecular-weight compounds that can be bound by immunoglobulins but stimulate no immune response—they are antigens that have no immunogenic properties. When haptens combine with select larger molecules, the immune system may respond to the joined complex much more strongly than it would to either the molecule or the hapten singly. Often, after the first exposure to a complexed hapten, the immune system will subsequently respond to the hapten alone or the molecule alone. In this way, the immune system is effectively "trained" to recognize the hapten as an immunogen, rather than only an antigen. Examples of haptens include some drugs, industrial chemicals such as dinitrophenol, and poison ivy toxin.

6.4 CELL-MEDIATED IMMUNITY

6.4.1 T Cell Subpopulations and Functions

Cell-mediated immunity is primarily mediated by T cells. T cells detect the presence of foreign substances (antigens) when these substances interact with T cell receptors on the cell membrane. T cell receptors are closely related to immunoglobulins, which helps explain the specificity and tight binding of T cell receptors to antigens. T cells also express a number of characteristic cell-surface proteins, which are classified by the abbreviation CD (which stands for "cluster of differentiation") and a number. For example, in mature cells, T cell receptors are generally expressed in conjunction with CD3 proteins because CD3 proteins

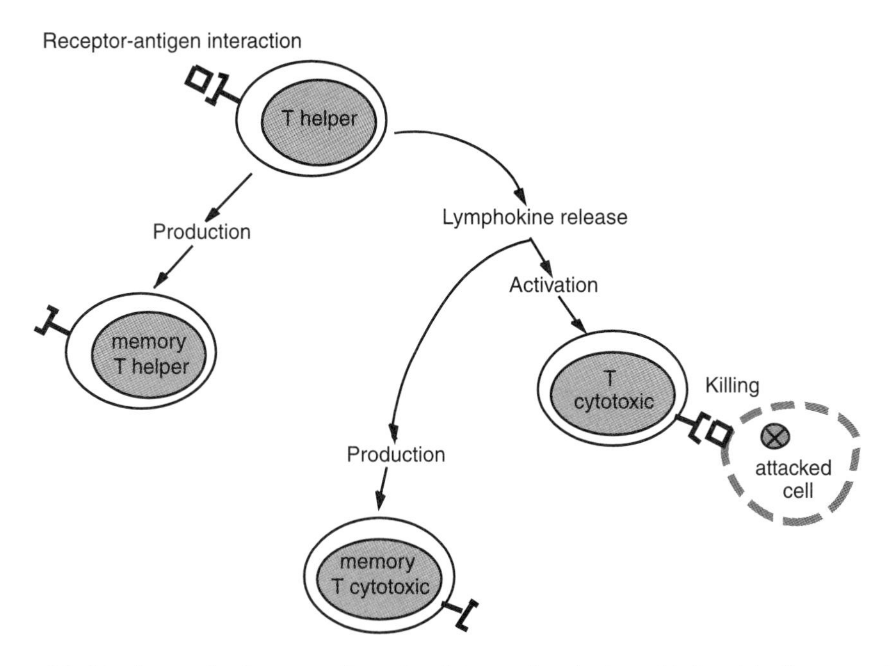

Figure 6.5. T cell contributions to cell-mediated immunity. Activated helper T cells secrete lymphokines and produce memory helper T cells. The lymphokines activate cytotoxic T cells and enable direct attack on foreign cells or objects that display the antigen stimulus. The concurrent production of memory cytotoxic T cells allows a quick and strong response to subsequent activations by the antigen stimulus.

help transmit signals from the T cell receptors into the cytoplasm. A cell that expresses CD3 proteins is designated as CD3+. There are a number of CD proteins, expressed in multiple combinations by various subpopulations of T cells at different developmental stages and with different functions. Most mature T cells are CD2+CD3+CD7+; T cells that additionally express CD8 tend to be *cytotoxic T cells*, able to directly attack and kill cells that express stimulatory antigens. T cells that do not express CD8 but instead express CD4 are generally *helper T cells*, which secrete chemicals (lymphokines) that help regulate the body's immune response. The ratio of CD4 to CD8 T cells in a healthy person is 2:1.

Figure 6.5 illustrates the interactions between helper and cytotoxic T cells. When a helper T cell is stimulated by contact with an antigen, the cell secretes lymphokines (for example, the *interleukins* and *granulocyte-monocyte colony stimulating factor*). Lymphokines cause B cells to proliferate and initiate the humoral immunity process (covered in Section 6.5); lymphokines also stimulate the accumulation of macrophages in an area and increase macrophage phagocytosis. Additionally, lymphokines (especially interleukin-2) stimulate the proliferation of cytotoxic T cells. Cytotoxic T cells bind to other cells that display antigens to the T cell receptors, and then the T cells secrete proteins called perforins, which create holes in the membrane of the attacked cell. The cytotoxic T cells then secrete potent toxins directly into the cytoplasm of the offending cell, which usually dies and dissolves quickly. The T cell receptors then disengage from the antigens on the surface of the attacked cell, which allows the T cell to move on and continue destroying antigen-expressing cells. Some of the cytotoxic T cells and some of the helper T cells, once stimulated by T cell receptor-antigen interactions, may differentiate into *memory T cells.* These cells can remain quiescent in the body for long periods of time (on the order of years), allowing the body to respond quickly and strongly to any subsequent exposure to the stimulatory antigens.

Cytotoxic T cells are efficient and effective little killing machines, and they are very important in the body's defense against viral infections because antigenic virus particles often become trapped in the membranes of infected cells, "marking" the infected cells for destruction. Cytotoxic T cells are also important in "self-recognition," destroying cells (from a tissue graft, for example) that display antigens not normally present in an individual patient. However, the importance of helper T cells should not be understated. Without the lymphokines secreted by helper T cells, the functions of the B cells and the proliferation of the cytotoxic T cells are severely compromised, nearly disabling the immune system. Helper T cells are destroyed in acquired immunodeficiency syndrome (AIDS), leaving the immune system unable to effectively combat many infectious diseases.

6.4.2 Antigen-Presenting Cells

T cells only recognize and respond to antigens after they have been processed and expressed in conjunction with a type of protein called *major histocompatibility complex* (MHC) proteins on the surface of *antigen-presenting cells.* Although there are multiple types of antigen-presenting cells, one type relevant to this textbook is the macrophage. Macrophages phagocytose a wide variety of foreign objects, and although the exact antigen-processing mechanisms are largely unknown, it appears that within the phagosomes of a macrophage, proteins are broken down into short peptides. Some of these peptides then bind to *class II MHC proteins* and are expressed on the membrane of the antigen-presenting macrophage (Fig. 6.6). Only helper T cells that possess the correct kind of T cell receptor can then bind to the complex of processed antigen and displayed MHC proteins, resulting in activation of the helper T cell and initiating an immune response.

Class II MHC proteins are mainly expressed by macrophages, and CD4+ T cells (helper T cells) only recognize antigens that are bound to class II MHC proteins. CD8+ (cytotoxic) T cells only recognize antigens when they are associated with *class I MHC proteins.* Fortunately, almost all cells in the body express class I MHC proteins on their membranes, so nearly any cell that expresses a foreign antigen can be recognized and attacked by cytotoxic T cells.

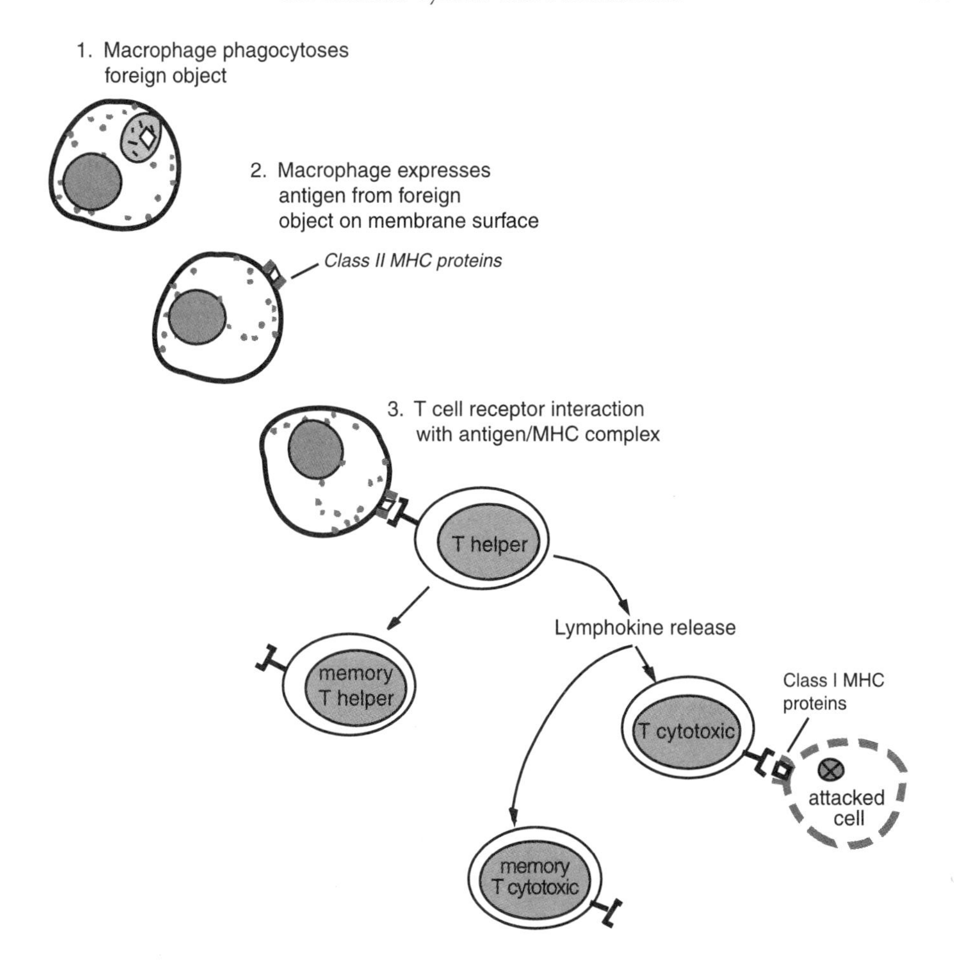

Figure 6.6. Antigen-presenting cell contributions to cell-mediated immunity. The antigen-presenting cell, in this case a macrophage, chemically digests a foreign object to break its structure down into small pieces (digesting a protein into short peptide fragments, for example). The small pieces of the immunogen are then expressed on the macrophage surface, in conjunction with class II MHC proteins. The helper T cell (CD4+) is only able to recognize and interact with the exposed antigen when it is situated with the class II MHC proteins. The helper T cell is activated by receptor engagement, and the events described in Figure 6.3 occur. The cytotoxic T cell (CD8+) is able to attach to (and destroy) cells that display the stimulus antigen in conjunction with class I MHC proteins.

6.5 HUMORAL IMMUNITY

6.5.1 B Cell Subpopulations and Functions

B cells can be activated via lymphokines secreted from helper T cells, notably interleukin-2, interleukin-4, and interleukin-6. B cells can also be activated by direct contact of B cell membrane immunoglobulin receptors with an antigen; unlike T cells, B cells do not require an antigen to be complexed with MHC

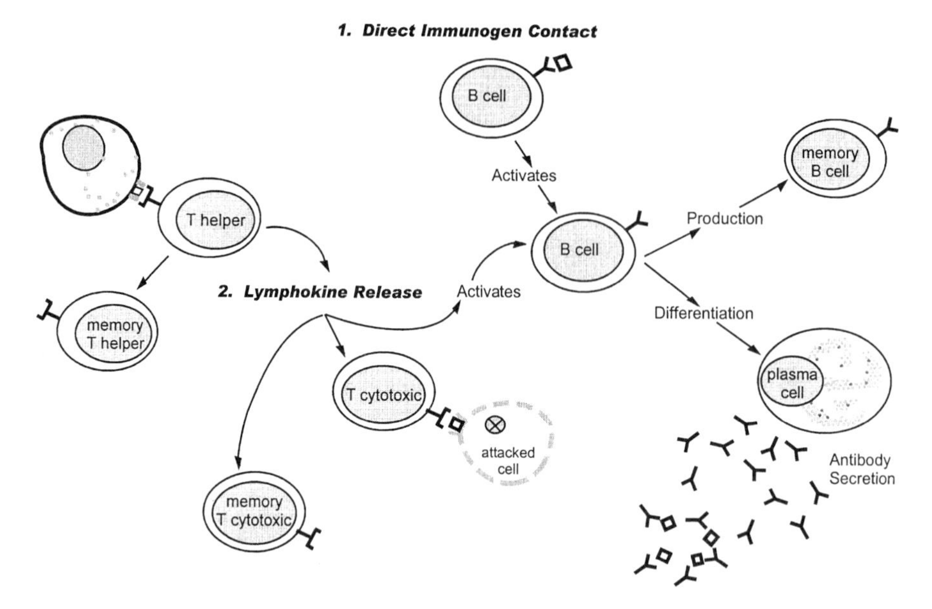

Figure 6.7. B cell contributions to humoral immunity. B cells can be activated in two ways: through direct contact and immunoglobulin receptor engagement with an antigen or from the lymphokines secreted by activated helper T cells. In either case, memory B cells are formed to prepare for any subsequent exposure to the stimulus antigen, and the stimulated B cells differentiate into plasma cells. The plasma cells secrete antibodies able to bind to the stimulus antigen.

proteins for recognition and engagement. Once B cells are activated, they go through a differentiation process to produce plasma cells (Fig. 6.7). Plasma cells are large cells with a well-developed rough endoplasmic reticulum, and mature plasma cells secrete large quantities of antibodies. It has been reported that a plasma cell can secrete as many as 2,000 antibody molecules per second, and that this rapid secretion can continue for several days or even weeks (Guyton, 1991). The secreted antibodies, which possess the appropriate variable portions to allow binding to the type of antigen that caused the initial immune stimulus, are released into the bloodstream. Some of the activated B cells differentiate into *memory B cells*, which can remain quiescent in the body for long periods of time (on the order of years). The memory B cells ensure that if the initial antigen stimulus is encountered again in the future, the body will have a nearly immediate source of appropriate antibodies.

Antibodies can inactivate antigens by covering the toxic or damaging regions of the antigen, effectively neutralizing the antigen. Antibodies can also deactivate antigens via agglutination, in which large antigens with multiple epitopes are bound together in a clump by antibodies and are thus unable to function. Antibody-antigen binding is so efficient that agglutination complexes can become large enough to precipitate an otherwise soluble antigen out of solution, inactivating the antigen. The protective effects of direct antibody-antigen contact are greatly amplified by activation of the *complement system.*

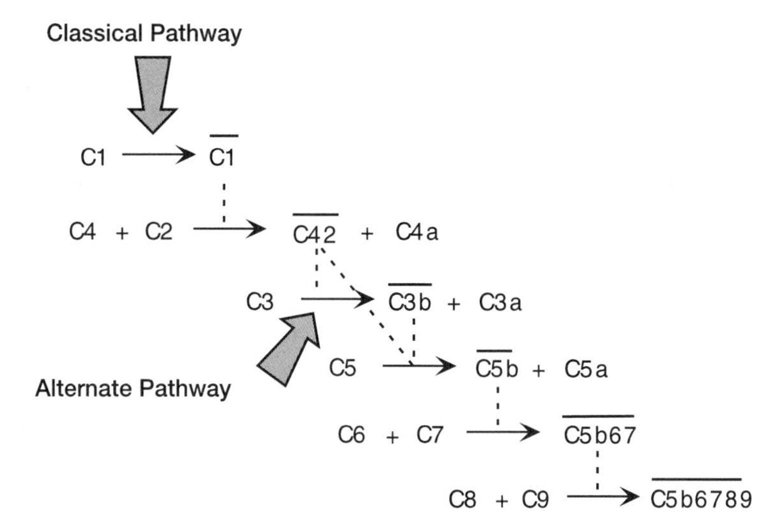

Figure 6.8. The complement cascade. Reaction products that are used in subsequent activation steps are designated with overhanging horizontal lines. Dashed lines indicate which subsequent reaction steps are affected by each reaction product. Information redrawn from A.C. Guyton, *Textbook of Medical Physiology*, eighth edition, W.B. Saunders Company, Harcourt Brace Jovanovich, Inc., Philadelphia, PA (1991), page 379.

6.5.2 The Complement System

The complement system is a cascade of activated enzymes, somewhat similar to the coagulation cascades discussed in Chapter 4. Complement proteins circulate in the plasma and can be activated by two major pathways. The classical pathway activates the complement system via an antibody-antigen reaction. When an antibody binds to an antigen, a site on the constant portion of the antibody changes conformation such that this site is able to bind the C1 molecule, setting off the complement cascade (Fig. 6.8). The alternate pathway begins when large polysaccharides from the membranes of invading microorganisms combine with complement proteins B and D; the resulting complex activates complement protein C3, and the cascade in Figure 6.8 continues. The alternate pathway can be initiated almost instantaneously on the entrance of a microorganism into the body, and so C3 activation and the complement system is part of the body's first line of defense against microorganisms.

The complement system is truly a cascade because with each reaction step, increasing quantities of products are formed. A number of the products of the complement system play important physiological roles. The terminal product shown in Figure 6.8, C5b6789, is called the membrane attack complex (MAC) and is capable of directly lysing the cell membranes of bacteria or foreign cells. Complement products can cause agglutination of foreign particles (similar to antibody-mediated agglutination) and can neutralize some viruses by interfering with the virus structure. C3b enhances macrophage and neutrophil phagocytosis

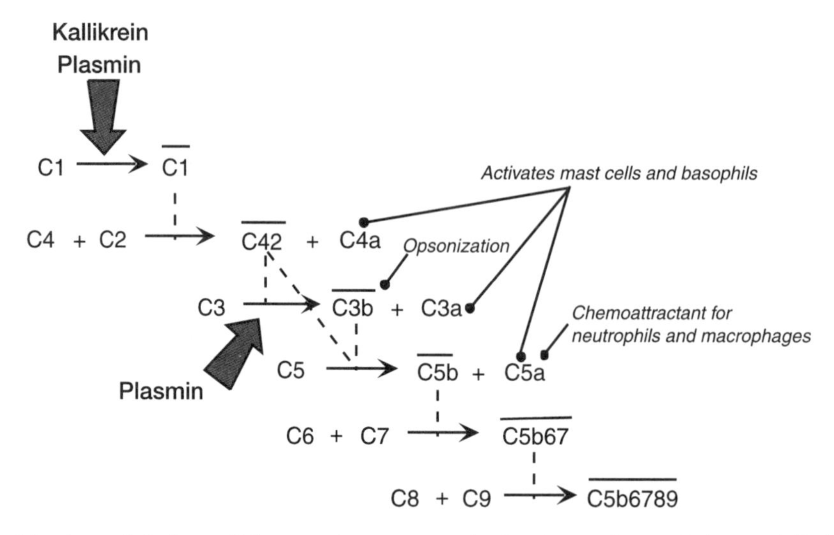

Figure 6.9. Amended chart of the complement cascade. Products of coagulation and fibrinolysis can activate the complement cascade. Given the various activation methods and some of the noted properties of complement products, the complement system constitutes an important link between coagulation and inflammation.

of foreign objects through opsonization. C3a, C4a, and C5a activate mast cells and basophils, and C5a is also a chemoattractant for neutrophils and macrophages. The actions of C3a, C3b, C4a, and C5a provide direct links between the inflammation process and the complement system. Moreover, by-products of the coagulation process—namely kallikrein and plasmin—are able to activate the complement system through the classical and alternate pathways (Fig. 6.9). The complement system therefore provides yet another link between the coagulation and inflammation stages of the wound healing process.

6.6 GENERATING SPECIFICITY

6.6.1 Clonal Selection Theory

It is hard to imagine how the immune system can specifically recognize and respond to so many antigens, and even more specific epitopes. The *clonal selection theory*, which has gained such wide acceptance that it is sometimes accepted more as a fact than as a theory, provides an explanation for how lymphocytes can be generated for such specific binding. The clonal selection theory assumes that the body is continually generating new lymphocytes, each of which expresses on its surface either immunoglobulin antibodies (B cells) or T cell receptors (T cells). The clonal selection theory further assumes that all of the lymphocyte surface receptors or antibodies on any given cell have the same chemical binding specificity (all the surface receptors or antibodies on any given cell will bind the same epitope) and that only a small subset of the entire lymphocyte pool will be able to bind to any given epitope. These assumptions are not inconceivable, because when lymphocytes proliferate outside of the thymus or bone marrow they essentially form clones of the original mother lymphocyte, identical in T cell or immunoglobulin receptor binding characteristics. Moreover, it has been estimated that approximately 10^9 newly formed "virgin" lymphocytes are released from the marrow and thymus into the blood stream each day, and it has also been estimated that the lymphopoietic system can produce lymphocytes with approximately 10^8 distinct specificities. Therefore, it is conceivable that on any given day, at least a few lymphocytes will express any given antibody or receptor type. The clonal selection theory predicts that only lymphocytes stimulated by antibody-antigen or T cell receptor-antigen interactions will proliferate and form memory cells; unstimulated (and thus unneeded) lymphocytes eventually die (Fig. 6.10). The incredible diversity of lymphocyte specificities implied by clonal selection theory is thought to result from the activities of multiple germ line genes and variations/mutations in how those genes are expressed.

6.6.2 "Self" Versus "Non-self"?

Clonal selection theory may explain how lymphocytes are able to recognize and interact with a wide variety of antigens, but how does a person's immune system "know" not to attack healthy cells or tissue from his or her own body? If lymphocytes can theoretically have 10^8 different specificities, then isn't it probable that antigens to at least some of those 10^8 variations are naturally expressed in a person's own body somewhere? How do lymphocytes differentiate between "self" and "non-self"?

The short answer is that, currently, scientists are not completely sure how the immune system discerns "self" from "non-self." It seems that some self-tolerance is developed in the womb: while embryonic tissues and cells are maturing, the developing immune system "accepts" any biomolecules, cells, and tissues as "self." Because of this, maternal IgG antibodies are able to cross the placenta and circulate in the bloodstream of a fetus, protecting the baby from infection. After birth, it seems that self-tolerance is generated by rigorous selection processes during lymphocyte maturation in the thymus or the bone marrow. Only T cells that recognize the exact class I and class II MHC proteins carried by the host survive in the thymus or marrow to begin the maturation process. This is termed *positive selection*, because it ensures that any mature lymphocytes will be able to respond appropriately to antigen-presenting cells of the host. As T or B cells mature in the marrow and thymus, they are surrounded by potential self-antigens. Any immature T or B cells that engage self-antigens and become activated undergo *apoptosis*, or programmed cell death. The trigger for this apoptosis appears to be that these defective T or B cells become activated too early—before maturity and thus before leaving the marrow or thymus—when the only antigens these immature cells should be exposed to are

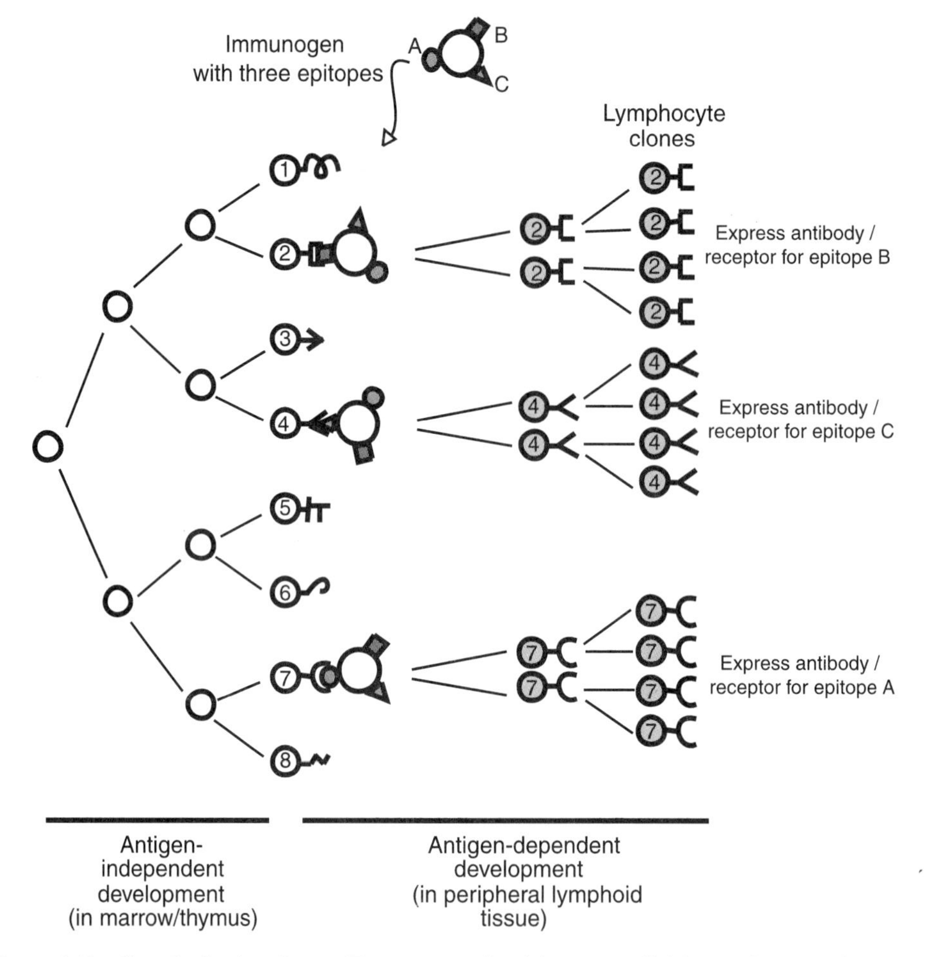

Figure 6.10. Clonal selection theory. From stem cell origins, some division and maturation occurs in the bone marrow or thymus, where "non-self" antigens should not be present. Lymphocytes in the marrow and thymus do not depend on antigen stimuli to replicate. In the peripheral lymphoid tissues, only the lymphocytes that are stimulated by engagement with an immunogen will clonally replicate. Thus only the lymphocytes that express "necessary" antibodies/receptors reproduce to combat the immunogen. Adapted and redrawn from Darnell, J., Lodish, H., and Baltimore, D., *Molecular Cell Biology*, second edition, W.H. Freeman and Company, New York, NY (1990), page 1011.

self-antigens. Automatic diversion to programmed death is a *negative selection* process because it reduces the chance that mature lymphocytes will respond inappropriately to self-antigens. Although some autoreactive T cells may escape the thymus, these cells tend to quickly become *anergic*, or quiescent and inactive, through mechanisms not yet fully understood. When lymphocytes lose the ability to discern between self and non-self antigens, and begin to attack the body's normal tissues, a variety of *autoimmune* disorders can result.

The body's ability to recognize and attack "non-self" antigens is the reason why transplanting organs and tissues from one individual to another requires careful preplanning and *tissue typing* and usually requires long-term medication to suppress some of the activity of the immune system. Tissue typing is essentially a series of tests to determine potential matches between organ/tissue donors and recipients once these individuals are known to possess compatible ABO blood types. The first step is to determine the types of human leukocyte antigens, or *HLA*, expressed by the potential donor and the patient. The HLA are also known as the major histocompatibility complex—the same MHC that was discussed previously in this chapter to describe the functions of antigen-presenting cells. Genes for the HLA/MHC complex are located on the short arm of human chromosome 6, and these genes code for the specific types of class I and class II MHC proteins, as well as for variations within the subtypes of class I and class II MHC proteins. Even though the genes that code for the HLA/MHC complex are inherited from an individual's parents, there is no guarantee that any two siblings (other than identical twins) will have the same sets of HLA antigens—will have the same amino acid sequences for their class I and class II MHC proteins. The greater the match between the HLA of a potential organ/tissue donor and the HLA of the recipient, the better the *histocompatibility* of the two individuals. If it is determined that the potential donor and recipient have an acceptable level of HLA matching, a sample of the potential donor's lymphocytes are exposed to a sample of the patient's blood serum. This *cross-matching* tests whether the serum contains antibodies to the donor's antigens, indicated by damage to the serum-exposed lymphocytes. Finally, *antibody screening* tests identify the amount of antibodies to HLA that are in the patient's blood serum, as well as the specific types of HLA antibodies in the serum. Although many of the tests used for tissue typing can be performed with living cells and blood serum from patients and organ/tissue donors, evolving molecular biology and genetic screening techniques can be used as well, usually providing a great deal of needed information in a short period of time. Even with careful screening and acceptable matches between organ/tissue donors and recipients, after a transplant occurs, the recipient must take immunosuppressive drugs to reduce the activity of the immune system and the chance that the organ/tissue will be recognized as foreign, attacked, and rejected. These drugs, many of which can have significant, unpleasant, or medically-difficult side effects, are expensive and often must be taken for the rest of the patient's life.

6.7 SUMMARY

Knowing the general mechanisms of B cell and T cell function is important to understanding why great care must be taken when implanting tissues or cells from one species into another, or even from one person into another. Tissue-engineered products that incorporate living cells or products of cells that were not derived from the intended patient must be carefully screened and processed so that significantly adverse immune reactions do not occur. And yet, the body's immune system can be a beneficial participant in the inflammatory stage of the normal process of wound healing. Lymphokines released by activated B cells can be chemoattractants for neutrophils and eosinophils, and cause mast cells and basophils to release histamine into tissues. Products of the complement cascade cause chemotaxis of neutrophils and macrophages and also activate mast cells and basophils. Several products of the complement system cause increased capillary permeability and vasodilation and thus swelling, heat, and pain. The processes of coagulation and inflammation are designed to quickly stop any loss of blood, slow the advance of any invading microorganisms, attract phagocytes to the area to remove damaged tissues, and attract lymphocytes to the area to destroy any bacteria, viruses, etc. that may infiltrate the wound.

At this point we can assemble and summarize the material from Chapters 4, 5 and 6 to reiterate a concept crucially important to understanding tissue-biomaterial interactions: *The processes of coagulation and inflammation can be triggered merely by the presence of a biomaterial surface that allows adsorption and activation of factor XII (Fig. 6.11).* Any material placed in the body will

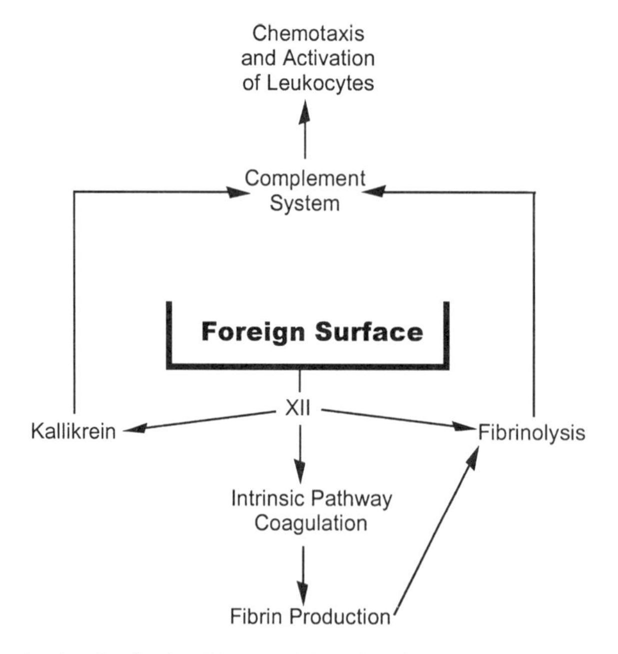

Figure 6.11. Central role of a foreign biomaterial surface in the initiation of coagulation and inflammation. Simply through blood-foreign surface (i.e., a surface that is not the normal, healthy endothelium) contact, factor XII will adsorb to the surface and initiate (either directly or indirectly) a variety of cascading reactions including the intrinsic coagulation pathway, fibrinolysis, complement activation, and the physiological bases for the cardinal signs of inflammation.

cause a biological response, as will the wound created during the material implantation procedure. The degree of response to an implanted biomaterial may be altered by carefully designing the physical and chemical characteristics of the biomaterial (discussed in Chapter 8), but to date, there is no such thing as an "inert" biomaterial.

6.8 BIBLIOGRAPHY/SUGGESTED READING

Abbas, A.K., Lichtman, A.H., and Pober, J.S., *Cellular & Molecular Immunology*, fourth edition, W.B. Saunders Company, Philadelphia, PA (2000).

Darnell, J., Lodish, H., and Baltimore, D., "Chapter 26: Immunity" in *Molecular Cell Biology*, second edition, W.H. Freeman and Company, New York, NY (1990), pp. 1003–1048.

Goldsby, R.A., Kindt, T.J., and Osborne, B.A., *Kuby Immunology*, fourth edition, W.H. Freeman & Company, New York, NY (2000).

Goodman, J.W., "Chapter 3: The Immune Response," in *Basic & Clinical Immunology*, eighth edition (Stites, D.P., Terr, A.I. and Parslow, T.G., eds.), Appleton & Lange, Norwalk, CT (1994a), pp. 40–49.

Goodman, J.W., "Chapter 4: Immunogens and Antigens," in *Basic & Clinical Immunology*, eighth edition (Stites, D.P., Terr, A.I. and Parslow, T.G., eds.), Appleton & Lange, Norwalk, CT (1994b), pp. 50–57.

Guyton, A.C., "Chapter 34: Resistance of the Body to Infection: II. Immunity and Allergy," in *Textbook of Medical Physiology*, eighth edition, W.B. Saunders Company, Harcourt Brace Jovanovich, Inc., Philadelphia, PA (1991), pp. 374–384.

Junquiera, L.C., Carneiro, J., and Kelley, R.O., "Chapter 14: The Immune System & Lymphoid Organs," in *Basic Histology*, eighth edition, Appleton & Lange, Norwalk, CT (1995), pp. 247–251.

Parslow, T.G., "Chapter 2: Lymphocytes & Lymphoid Tissues," in *Basic & Clinical Immunology*, eighth edition (Stites, D.P., Terr, A.I. and Parslow, T.G., eds.), Appleton & Lange, Norwalk, CT (1994), pp. 22–39.

Roitt, I.M., *Roitt's Essential Immunology*, Blackwell Scientific Incorporated, Boston, MA (1997).

6.9 STUDY QUESTIONS

1. Explain the differences between an immunogen and a pathogen; an antigen and an immunogen; an epitope and an antigen.
2. Relate the structure of immunoglobulins to their epitope-specific binding and to the specificities of T cells and B cells.

3. Predict possible consequences of the introduction of haptens into the body.
4. Explain how the immune system retains "memories" of previously encountered antigens.
5. Make your own diagram that explains, in detail, the activation and functions of helper and cytotoxic T cells, antigen-presenting cells, B cells, and plasma cells.
6. Discuss the specific roles of MHC proteins and antibodies in the immune system.
7. Describe the activation mechanisms and outcomes of the complement system as related to the processes of coagulation and inflammation and the presence of a "foreign" surface in the body.
8. Explain and justify the assumptions and implications of clonal selection theory.
9. Describe why lymphocytes generally do not react to self-antigens, and what can happen when they do.
10. Answer this question: "What exactly is tissue typing?"
11. Summarize major parameters that make biomolecules more or less immunogenic.

6.10 DISCOVERY ACTIVITIES

1. Some researchers are working on encapsulating living cells within various biomaterials, such that a given population of cells could be implanted into a number of patients and protected from destruction by the patients' immune systems. Do a literature search on encapsulating cells for immunocompatibility. What are some clinical problems that could be addressed with this technology? What types of biomaterials are being developed for this purpose? What scientific challenges have yet to be met?

2. Use the World Wide Web to discover how, in general, organs are obtained and distributed in this country. What is the United Network for Organ Sharing (UNOS)? What is the Department of Health and Human Services, and what does it have to do with UNOS? Choose one organ to focus on from this list: heart, kidney, liver, pancreas, intestine. How many people in the U.S. are currently waiting for a transplant of this type? How much does this type of transplant tend to cost, and how do people pay? What kinds of follow-up care/procedures are required? What are some sources of information for people who are awaiting or who have received this type of organ transplant?

3. Read and discuss Chapter 13, "Allocation of Artificial and Transplantable Organs," from G.E. Pence, *Classic Cases in Medical Ethics: Accounts of Cases that Have Shaped Medical Ethics, with Philosophical, Legal, and Historical Backgrounds*, third edition, The McGraw-Hill Companies (2000).

4. Read "The Danger Model: A Renewed Sense of Self," by P. Matzinger (*Science* 296:301–305, 2002), and compare the Danger model with the mechanisms of self/non-self recognition summarized in this text.

7

Wound Healing

7.1 INTRODUCTION

Tissue injury involving cell death, destruction of extracellular connective tissue components, and loss of blood vessel integrity automatically triggers the wound healing process. Wound healing, the natural response to injury, is a remarkable testament to the body's capacity to regenerate cells characteristic and pertinent to specific tissues and organs, replace connective tissues and blood vessels, and thus repair damage.

Wound healing, a complex (but, in healthy subjects, effective and efficient) system of checks and balances, involves a series of highly interdependent processes that overlap in time. The underlying tissue repair mechanisms (especially at the molecular level) are still, at best, partially understood; what has been established, however, is that normal wound healing is regulated by the balanced actions of growth factors, cytokines, proteases and other, presently unknown, chemical compounds and conditions. In addition, there is agreement that injury to tissue triggers a sequence of histologic changes and that the process of wound healing can be conceptualized as a sequence of functionally distinct phases that result in neotissue formation and thus to wound closure.

7.2 TISSUES

Tissues are integral and fundamental components of organs. Tissues consist of pertinent and specific cells (which are distinct from blood cells) and of extracellular matrix that is formed and maintained by chemical compounds (such as proteins) synthesized by cells. For example, bone consists of osteoblasts (the bone-forming cells), osteocytes, and osteoclasts (the bone-resorbing cells). These cells reside in a matrix that is a composite of an inorganic (approximately 65%) and an organic (approximately 30%) phase. The predominant components of the bone matrix are calcium-containing mineral in the inorganic phase but mostly collagen type I and small amounts of noncollagenous proteins in the organic phase. Endothelial cells, smooth muscle cells, and fibroblasts are the cells of the wall tissue of the large blood vessels; the respective extracellular matrix contains collagen type IV and elastin. Skin contains keratinocytes and epithelial cells in a matrix of collagen (types IV and VII) and elastin. Endothelial cells, the innermost layer of blood vessels, are present in all vascularized tissues; nerve cells are found in innervated tissue (such as bone, blood vessel wall, skin). Last, but not least, fibroblasts are ubiquitous and are found in many tissues. Other extracellular components are adhesive proteins (such as fibronectin, laminin, and vitronectin; see Section 7.3.2.2), proteoglycans, ions, and water.

Tissue injury may affect the morphology, function, and phenotype of cells. These effects may be reversible or irreversible, short-term or permanent, and may lead to cell death. Examples of cell responses to injury are atrophy (decrease in cell size and/or function), hypertrophy (increase in cell size), hyperplasia (increase in cell numbers), metaplasia (change of the cell type), and change of phenotype (that is, changes in the type and amount of production of proteins characteristic of a specific cell type). Clinical examples that illustrate such cell responses are bone loss in the vicinity of orthopedic metal implants due to stress-shielding and intima hyperplasia (due to enhanced proliferation of smooth muscle cells) at the anastomotic sites of vascular grafts attributed to mismatch of the compliances of the synthetic material and of the native blood vessel tissue (see Chapter 10, Example 1).

Another aspect that critically affects the outcome of wound healing in tissues is the regenerative capacity of various cells. For example, labile cells (such as epithelial, lymphoid, and hematopoietic cells) proliferate throughout life; stable cells (such as fibroblasts, smooth muscle cells, osteoblasts, chondrocytes, and vascular endonthelial cells) have the capacity to proliferate and do so in response to appropriate stimuli. In contrast, permanent or static cells (such as nerve, skeletal and cardiac muscle cells) cannot reproduce themselves after birth. Theoretically, then, tissue repair in response to injury caused by disease and surgery can be only expected in tissues that contain either labile or stable cells; in contrast, restitution in tissues with permanent cells is limited, if it occurs at all.

7.3 THE BIOLOGY OF WOUND HEALING OF VASCULARIZED CONNECTIVE TISSUE IN ADULTS

7.3.1 Pertinent Aspects of Hemostasis and Inflammation

Damage to blood vessels in vascularized tissue, and concomitant blood leakage, triggers blood coagulation. The blood clot (a fibrin mesh in which blood-derived cells are interspersed) at the site of injury plugs the defect and provides temporary protection to the exposed wound site as well as a provisional matrix for cells to grasp hold of (attach) and crawl over (migrate) during the healing process. Moreover, the processes of blood coagulation (described in Chapter 4) and activation of the complement cascade of chemical reactions (described in Chapter 5) initiate the inflammatory stage of wound healing.

Among the soluble chemical mediators released from cytoplasmic granules during activation and degranulation of platelets are growth factors. The terms "cytokines" and "growth factors" are often used interchangeably in the scientific literature. For the purpose of this textbook, the term growth factors will be used to describe peptides and proteins that stimulate cell differentiation, proliferation, migration, and other functions. Table 7.1 presents an overview of the sources and actions of some growth factors important to wound healing. For example, growth factors released from platelets (such as platelet–derived growth factor and transforming growth factor-beta) are potent chemoattractants for inflammatory leukocytes, as well as motogens and mitogens for cells specific to various tissues. Growth factors may either induce cell migration against a concentration gradient of the growth factor (*i.e.*, chemotaxis), or they may simply increase the rate of random, undirected cell migration. Moreover, growth factors produced by a given cell type can act in an *autocrine* manner (affecting the function of that cell type), or in a *paracrine* manner (affecting the function of other cell types in the surrounding tissues). A number of growth factors are released from platelets during coagulation and from leukocytes during inflammation; these growth factors may cause cells (fibroblasts, endothelial cells, etc.) from surrounding tissues to produce their own growth factors and thus help regulate wound healing.

Growth factors and cytokines are also released from activated leukocytes during inflammation. At the site of the damaged tissue, neutrophils and macrophages phagocytose and thus remove dead cells, extracellular matrix debris, and foreign bodies such as invading bacteria. Neutrophils release proinflammatory cytokines (such as interleukin-1, a chemoattractant for monocytes, and tumor necrosis factor) that activate local connective tissue cells. Activated macrophages release growth factors such as fibroblast growth factor, platelet-derived growth factor (both potent chemoattractants and mitogens for fibroblasts) and vascular endothelial growth factor (a potent angiogenic factor essential in the process of new capillary formation). In this fashion, the inflammatory leukocytes prepare the wound site and provide the chemical stimuli crucial for the next stage of healing, the proliferative phase.

7.3.2 Proliferative and Repair Phase

7.3.2.1 Cell Proliferation

Diffusion of chemoattractants (such as growth factors derived from activated platelets) from the wound into the surrounding tissues and body fluids rapidly attracts neutrophils, followed by monocytes, from the circulating blood in the vasculature to the wound site. The growth factors initially released by activated platelets are degraded by proteases through normal biochemical processes but are replaced by others (including epidermal growth factor, insulin-like growth

TABLE 7.1. Properties of Select Growth Factors Important to Wound Healing

Growth Factor	Produced By	General Functions
Epidermal Growth Factor (EGF)	Epidermal Cells	• Mitogenic for fibroblasts, smooth muscle cells, keratinocytes • Chemotactic for epithelial cells and fibroblasts • Important for re-epithelialization
Fibroblast Growth Factor (FGF) *Note:* While there are at least nine subtypes of FGF, this table refers only to acidic FGF (aFGF or FGF-1) and basic FGF (bFGF or FGF-2).	Endothelial Cells Macrophages	• Wide-spread mitogenic activity • Important mediator of angiogenesis and epithelialization • Modulates endothelial cell proliferation and migration • Increases fibroblast proliferation and production of extracellular matrix components
Insulin-Like Growth Factor (IGF) *Note:* There are two subtypes of IGF: IGF-I, and IGF-II.	Fibroblasts Smooth Muscle Cells Macrophages	• Stimulates fibroblast production of ECM components
Platelet-Derived Growth Factor (PDGF)	Platelets Macrophages Endothelial Cells Fibroblasts Smooth Muscle Cells	• Wide-spread mitogenic activity, especially for fibroblasts and smooth muscle cells • Chemotactic for neutrophils and macrophages • Considered to be a major stimulant of wound healing
Transforming Growth Factor-β (TGF-β) *Note:* There are at least five subtypes of TGF-β, designated TGF-β1 through TGF-β5. This table refers to TGF-β1.	Platelets Macrophages Endothelial Cells	• Regulates production of extracellular matrix components by fibroblasts • Inhibits degradation of newly-formed collagen • Inhibits proliferation of lymphocytes and some epithelial cells

Note. For more specific information and references relevant to the general information in this table, please see: Mutsaers, S.E., Bishop, J.E., McGrouther, G., Laurent, G.J. (1997) "Mechanisms of tissue repair: from wound healing to fibrosis," *Int. J. Biochem. Cell Biol.*, 29:5–17; and Declair, V. (1999). "The Importance of Growth Factors in Wound Healing," *Ostomy/Wound Management*, 45:64–80.

factor, fibroblast growth factor, and vascular endothelial growth factor) produced by neutrophils, macrophages, fibroblasts, epithelial, epidermal, and endothelial cells recruited to the wound site by the bioactivity of the growth factors initially released from platelets.

The signals received by cells from their surroundings (for example, upon contact with other cells, during interactions with the extracellular natrix, and by exposure to locally-released growth factors) activate and/or induce expression of cell-membrane receptors. Absence of neighbor cells at the wound site may signal both migration and proliferation of cells. Cell-cell and cell-matrix adhesion interactions are critical to the regulation of cell migration and proliferation in wound healing, and are mediated by various families of cell adhesion molecules (for example, integrins, cadherins, selectins, immunoglobulins, etc.). Among the cell adhesion receptors, integrins are important for various cell functions and are well studied; they are briefly reviewed in Section 7.3.2.1.1. For example, for cell migration to occur collagen receptors must be downregulated while integrins that bind to the adhesive ligand domains of fibrin, fibronectin, and vitronectin are upregulated. Proteases such as serum-derived plasmin (which breaks down the fibrin mesh of the blood clot) are available at the wound site. Collagenase and other matrix metalloproteinases (MMPs, each of which cleaves a specific subset of matrix proteins) are also upregulated in connective tissue cells at the wound-edge. These processes are prerequisites for cells to traverse the fibrin clot and remnants of the original extracellular matrix during their migratory phase.

Growth factors released at the wound site initiate mitosis of sedentary connective tissue cells at the wound margin. As a result, the wound site is "repopulated" with cells. Little is known about the signals involved in stopping cell differentiation and proliferation at this stage of wound healing; contact inhibition as well as other chemical (including desensitization receptors on the cell membrane) and mechanical cues may be involved. Consequently, the cells return to a quiescent state similar to that existing before injury.

7.3.2.1.1 Integrins

Integrin-mediated cell adhesion (and the intracellular signaling pathways that are triggered by this event) is important to a variety of physiologic and pathologic processes, including embryonic development, maintenance of tissue integrity, leukocyte circulation and recruitment, phagocytosis, and, most relevant to this chapter, wound healing and angiogenesis. Integrins are cell-membrane adhesion receptors for specific domains (ligands) of various proteins of the extracellular matrix, such as collagen, fibronectin, laminin, and vitronectin. Integrins are *heterodimeric* (having two different parts) structures and consist of an α subunit and a β subunit (Figure 7.1). Both subunits are glycoproteins, containing a large extracellular domain, a segment which spans the cell membrane, and a short cytoplasmic tail. The integrin family contains sixteen α and eight β subunits which can combine to form at least twenty-two distinct heterodimeric receptors. A given cell can express more than one type of integrin receptors, can modulate the number and spatial distribution of these receptors on its membrane, and thus affect cell interactions with ligands. A well-known and much-studied example of an integrin ligand is the short peptide Arginine-Glycine-Aspartic Acid, or RGD. The RGD sequence is present in a number of proteins

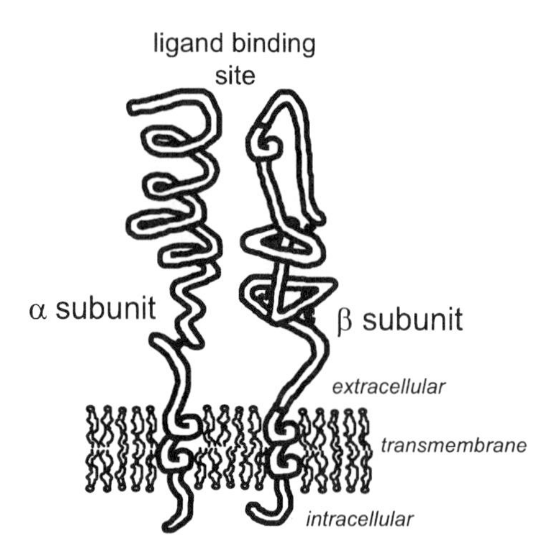

Figure 7.1. Schematic representation of an integrin receptor. Extracellular ligands bind at the tip of the α and β subunits. Divalent cations (such as Mg++) are required for integrins to bind to a ligand. Cations interact with receptors in the "pocket" created between the α and β subunits. Adapted from R.H. Kramer, J. Enenstein, N.S. Waleh, "Integrin structure and ligand specificity in cell-matrix interactions," in *Molecular and Cellular Aspects of Basement Membranes*, D.H. Rohrbach and R. Timpl (eds.), Academic Press, Inc., Harcourt Brace Jovanovich, New York, pp. 239–258 (1993).

important to the processes of wound healing, including von Willebrand Factor, fibrinogen, fibronectin, vitronectin, and collagen (see Chapter 2).

Integrin receptors at the *focal adhesions* points provide a direct link between the extracellular environment and the intracellular proteins of the cytoskeleton (Figure 7.2). The dynamic formation and disruption of focal adhesions is crucial for the process of cell migration. When integrin receptors bind ligands, a number of intracellular chemical events may occur including stimulation of protein kinase C and the Na^+/H^+ antiporter, phosphoinositide hydrolysis, tyrosine phosphorylation of membrane and intracellular proteins, changes in intracellular pH and calcium concentration, and mitogen-activated protein kinase activation. These chemical events initiate intracellular signaling pathways that are essentially lines of communication between integrins and the cell nucleus. The intracellular signaling that is initiated through integrin receptors binding to an extracellular ligand is often termed "outside-in" signaling, since the intracellular chemical events can be viewed as a response to 'signals' from the extracellular matrix. Integrins may also conduct "inside-out" signaling, since a cell can, by altering either the conformation of the cytoskeleton or expression/activation of integrin receptors, regulate its interactions with the extracellular milieu. Some of the intracellular signaling pathways that are "turned on" by integrin-ligand binding converge with pathways used by growth factors.

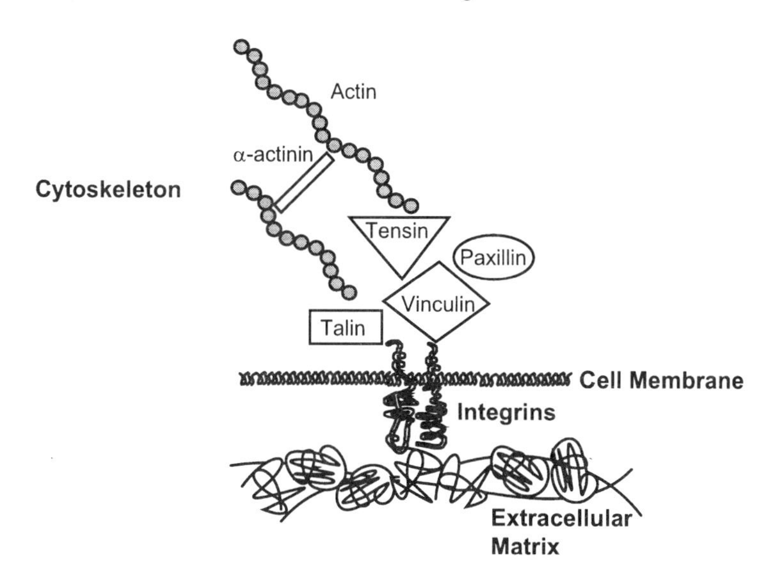

Figure 7.2. Schematic of a focal adhesion point. Integrin receptors constitute a direct link between the extracellular and the intracellular environments of the cell, and thus provide an "anchor" for the cytoskeleton.
Note: Only a few of the proteins of the intracellular network associated with focal adhesion points are shown in this schematic.

7.3.2.2 Extracellular Matrix Formation

Concomitantly with migration and proliferation, cells (primarily fibroblasts) synthesize extracellular matrix proteins that are deposited in an ordered sequence from the margin of the wound inward; a provisional extracellular matrix is thus formed. Table 7.2 lists and summarizes the functions of some extracellular matrix components that are important in various stages of wound healing.

Growth factors promote synthesis of extracellular matrix proteins by connective tissue cells. For example, in addition to stimulating migration and mitosis of epithelial cells, epidermal growth factor also promotes synthesis of fibronectin, one of the adhesive proteins of the extracellular matrix; transforming growth factor-β stimulates production of collagen, fibronectin, and glycosaminoglycans by fibroblasts; and platelet-derived growth factor stimulates production of fibronectin and hyaluronic acid by fibroblasts. Inhibitors [such as some cytokines (e.g., interleukin-1) and prostaglandins] of protein synthesis provide the controlling balance in this process.

The extracellular matrix is a dynamic milieu. It is the locus of growth factor accumulation; in fact, the extracellular matrix controls availability as well as bioactivity of growth factors at the healing wound site. Most important, it provides the substrate that anchorage-dependent cells need to firmly adhere; this event is prerequisite for subsequent cell functions (such as migration and syn-

TABLE 7.2. Extracellular matrix (ECM) components important to wound healing

ECM Component	Selected Functions
Collagens	Tissue structure and tensile strength Cell-ECM interactions Interactions between ECM components
Elastin	Tissue structure and elasticity
Fibronectin	Tissue structure Cell-ECM interactions Cell proliferation and migration Interactions between ECM components
Fibrinogen	Cell proliferation and migration
Hyaluronan	Cell-ECM interactions Cell proliferation and migration Interactions between ECM components
Laminin	Cell-ECM interactions, including migration
Proteoglycans	Cell-ECM interactions Cell proliferation Interactions between ECM components Binding growth factors
SPARC Also known as osteonectin	Cell-ECM interactions Anti-adhesive, anti-proliferative
Tenascin	Cell-ECM interactions Anti-adhesive and anti-proliferative function
Vitronectin	Cell-ECM interactions Interactions between ECM components Cell proliferation and migration

Adapted from Mutsaers, S.E., Bishop, J.E., McGrouther, G., Laurent, G.J. (1997). "Mechanisms of tissue repair: from wound healing to fibrosis," *Int. J. Biochem. Cell Biol.*, 29:5–17.

thesis of proteins) to occur. Cell interactions with their extracellular matrix may regulate cell function via receptor-mediated signaling mechanisms and/or by modulating cell responses to growth factors.

7.3.2.3 Neovascularization (Angiogenesis)

Formation of new blood vessels (by budding or sprouting of preexisting vessels) is necessary to sustain the newly formed tissue. Angiogenesis is a complex process that relies on the presence of an extracellular matrix in the wound bed as well as on migration and mitogenic stimulation of endothelial cells. Various growth factors (primarily fibroblast growth factor and vascular endothelial cell growth factor), released by macrophages and endothelial cells in the hypoxic and acidic environment of the wound site, stimulate growth of new capillaries into the neotissues, thus supplying the life-sustaining oxygen and nutrients contained in circulating blood.

Neovascularization involves proliferation, maturation, and organization of endothelial cells into capillary tubes. In addition to angiogenic growth factors, therefore, appropriate ligands (provided by the proteins in the provisional extracellular matrix) and receptors (such integrins) on the endothelial cell membrane are necessary for cell migration and other pertinent functions. Angiogenesis ceases once the wound site is filled with new tissue. Many of the new blood vessels disintegrate as a result of apoptosis (programmed cell death), a process regulated by a variety of matrix molecules.

7.3.2.4 Granulation Tissue

The new vascularized tissue contains many small blood vessels as well as cells and components (primarily proteins) that are pertinent to a specific tissue of the body. Numerous capillaries give this tissue (red in appearance, delicate, and prone to bleeding if disrupted) a granular appearance, hence the designation "granulation tissue." Macrophages are present in granulation tissue. Proliferating fibroblasts synthesize and deposit extracellular matrix components; in fact, the presence of proteoglycans and of collagen are indicative of the early and latter, respectively, stages of granulation tissue development. Overall, granulation tissue is histologic evidence of normal resolution of the inflammatory phase of the wound healing process

During the latter stages of granulation tissue development, the connective tissue is compacted and the wound contracts. Although the detailed mechanisms of these events are not fully understood, it is known that this phenomenon is the result of the contractile function of myofibroblasts (that is, fibroblasts that have assumed the smooth muscle cell phenotype and have the capacity to generate strong contractile forces). Wound contraction involves interactions of chemical compounds (notably cytokines), cells, and extracellular matrix. Tugging on myofibroblasts among themselves and on the surrounding matrix draws the wound margins together and closes the wound. Apoptosis of myofibroblasts may occur at the end of wound contraction.

7.3.3 Remodeling Phase

At the latter stages of damaged tissue replacement by neotissue (specifically, when abundant collagen has been deposited at the wound site), the cellularity of the wound decreases as a result of cells undergoing apoptosis triggered by unknown signals. A relatively acellular scar thus replaces the granulation tissue. Concomitantly, the extracellular matrix is remodeled via processes that involve changes in collagen synthesis and degradation; these events affect the relative amounts of various collagens (characteristic of specific tissues) as well as structural characteristics (such as thickness and orientation) of the collagen fibrils. Several proteolytic enzymes (such as matrix metalloproteinases) secreted by various cells (macrophages, epidermal cells, endothelial cells, and fibroblasts) control the degradation of collagen (catabolism) in the wound site. Even when the wound healing process is completed without complications, the resulting new tissue may not have the functional characteristics (such as structure, chemical composition, and mechanical properties) of the original tissue. Remodeling, therefore, may lead to scar tissue formation.

7.3.4 Scar Tissue

Because of changes in the relative amounts, type, and structure of collagens, scar tissue composition and structure are different from those of normal tissues. Moreover, because of collagen arrangement in dense parallel bundles, scars have only fractions of the strength of normal tissue; for example, uninjured skin characterized by a mechanically efficient basket-weave arrangement of collagen in the extracellular matrix. Scar tissues are also areas prone to reinjury. The underlying mechanisms of scar formation are not fully understood; there is evidence, however, that in addition to chemical factors (for example, high levels of transforming growth factor-β for the duration of wound healing), mechanical stresses at the wound site may affect collagen remodeling.

A unique aspect of abnormal wound healing concerns the aesthetic effectiveness of the wound healing outcome; a clinical example in this case are fibroproliferative disorders (such as keloids), which are hypertrophic scars with excess accumulation of collagen bundles in random orientation. Scars of skin wounds in visible regions of the body (for example, the face, hands, and arms) are undesirable outcomes that affect the quality of life of many patients. Subsequent plastic surgery to remove such scars may initiate another wound healing cascade with a similar final outcome, namely, new scars. For these reasons, clinical treatment of wounds focuses on timely and uncomplicated wound closure as well as a functional and aesthetically satisfactory scars.

7.3.5 Timing Aspects

A critical aspect for the success of wound healing is progression through the sequence of all stages of the process in a "timely" fashion. It should be noted that there is considerable overlap of the various stages of wound healing and that the time periods associated with each stage may vary because of species, extent of injury, and other medical conditions and complications.

After injury, neutrophils rapidly migrate and predominate at the wound site during the first 1–2 days. Monocyte migration and macrophage accumulation continue for days to weeks. Macrophages and their function are pivotal for the transition from the inflammatory to the proliferative stage of wound healing. Granulation tissue appears within 3–5 days, and angiogenesis takes place within the first 3–7 days after injury. The proliferative/repair phase lasts for several weeks, whereas the remodeling phase can last up to a couple of years.

7.3.6 Factors that Affect the Wound Healing Outcome

Progress and outcome of the wound healing process are affected by various factors including the proliferative capacity of surrounding cells that are pertinent (parenchymal) to the specific tissue/organ (that is, the site of the wound), the severity of injury and extent of tissue destruction, and the species, health condition, and age of the subject.

7.3.7 An Example of Wound Healing Tissue: Skin

Most current knowledge on wound healing has resulted from clinical treatment and relevant research of skin wounds in adults. In skin, the stages of healing can be monitored by visually examining the site of small wounds. A yellow wound is indicative either of the inflammatory stage or of infection. A red wound site (due to numerous new small blood vessels) is evidence of granulation tissue (the wound "connective tissue"). A shiny, pink wound site is one undergoing re-epithelialization (a process that involves proliferation and migration of epithelial cells from the edge toward the center of the wound, thus covering the wound). A wound with a black base is evidence of necrosis (dead tissue) either because healing has not been achieved or because the injury is chronic. In the case of large wounds, all stages of healing could be present at different regions of the wound at one time. Under normal circumstances, most skin wounds heal within a week or two after injury.

The rate at which healed skin tissue gains tensile strength is slow and parallels collagen accumulation and remodeling. Healed wounds never attain the same breaking strength (the tension at which skin breaks) as uninjured skin; the maximal strength of scars is only 70% that of normal skin.

7.4 CHRONIC NONHEALING WOUNDS

Defective (in most cases of unknown etiology) healing processes are associated with chronic nonhealing wounds, that is, wounds that either failed to heal or are healing at an abnormally slow rate. Diabetic ulcers are one example of impaired wound healing.

In the case of skin these are wounds in which healing was initiated but was not resolved through the various stages of the healing process to achieve wound closure. In normal healing the inflammatory response resolves as granulation tissue forms. In contrast, chronic infection, for example, may arrest the wound in the inflammatory stage (in which activated blood-derived inflammatory leukocytes and their chemical secretions control the process) and may never reach resolution of healing. Although the underlying mechanisms are not known, impaired matrix formation, inadequate or inactive growth factors, and misregulated enzymes (such as matrix metalloproteinases) may contribute to clinical non-healing syndromes. Chronic nonhealing wounds are accompanied by pain and suffering of patients and are associated with high health care costs.

7.5 WOUND HEALING AROUND IMPLANTS IN ADULTS

The stages of wound healing can be monitored (for example, by visual inspection) after skin injury. Evaluation of the wound healing process around implants, however, cannot be accomplished either easily or directly. It is generally accepted that, the overall, the normal, physiological scenario is applicable, with several distinct differences and variations on this theme. For example, the presence of neutrophils is evidence of acute inflammation (Fig. 7.3) whereas the presence of monocytes/macrophages, lymphocytes, and plasma cells in histologic specimens of tissues surrounding implants is evidence of chronic inflammation (discussed in Chapters 5 and 6). Foreign body reaction and formation of granulation tissue is the normal wound healing response to biomaterials/devices implanted in the body.

Implant devices are placed inside the body via invasive surgical procedures that cause cell, tissue, and organ injury and thus trigger the wound healing response of surrounding vascularized tissue. In addition to other factors (mentioned in Section 7.3.6), the geometry (that is, shape and size) of implants as well as the physicochemical properties of constituent biomaterials may affect the duration and intensity of the various steps of the physiologic wound healing process. Biomaterials adsorb serum-derived opsonins (such as immunoglobulins and activated complement protein fragments) that mediate subsequent adhesion and activation of neutrophils and macrophages. When implants are involved, however, the process of phagocytosis is incomplete (frustrated phagocytosis). Enzymes and oxygen radicals are synthesized and released by adherent, activated inflammatory leukocytes, but engulfment of large implants by the much smaller cells is not possible (Fig. 7.4).

In the attempt to phagocytose the synthetic material, monocytes and macrophages fuse and form multinucleated "foreign body giant cells." Histologic examination has provided evidence that macrophages and/or foreign body giant cells are present at the tissue-implant interface and may persist for the duration of implant residence in the body of recipient human and animal subjects. Many more macrophages and foreign body giant cells are found around textured material surfaces and on implants with high surface-to-volume ratios than on smooth surfaces. The presence and function of macrophages and/or foreign body giant cells at the implantation site during wound healing constitutes the

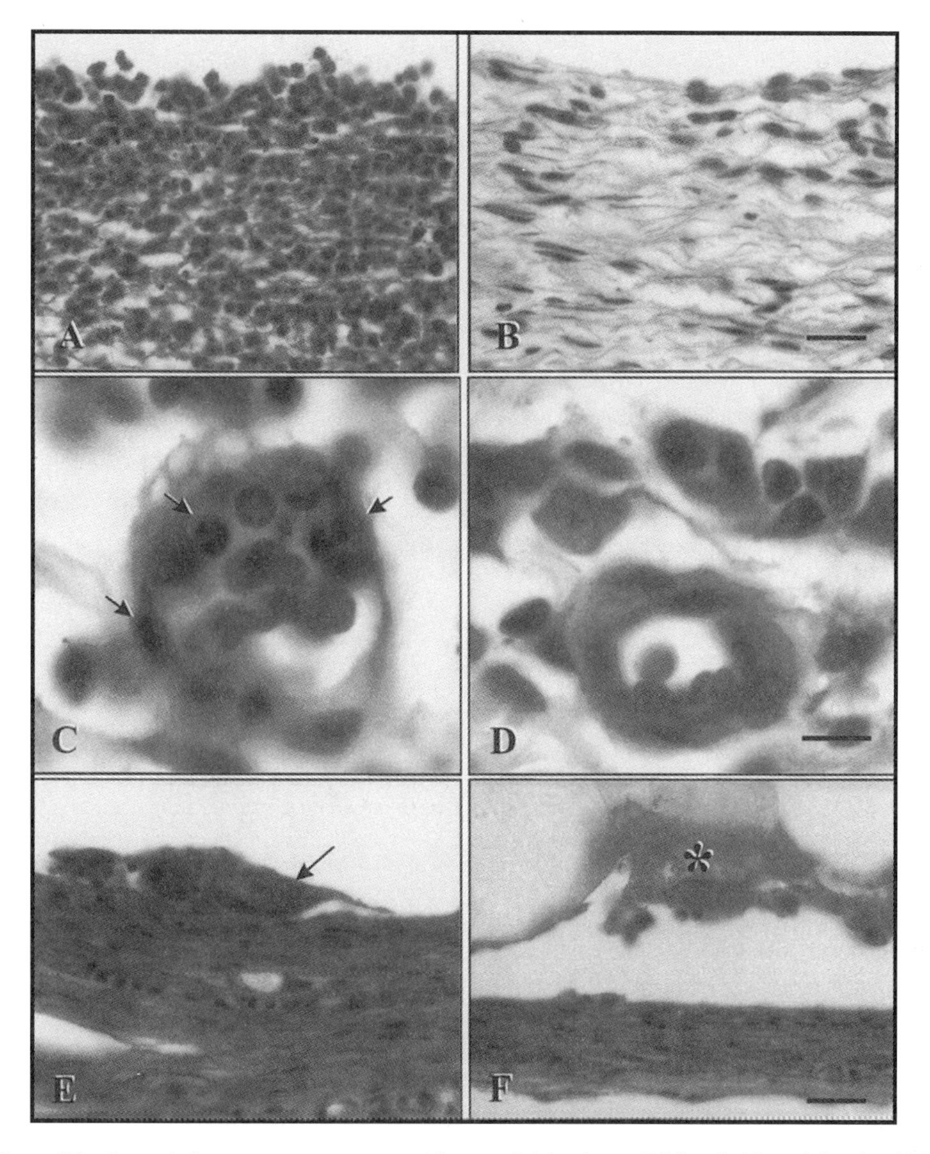

Figure 7.3. Acute inflammatory response to biomaterial implants. Disks of either plain ultrahigh-molecular-weight polyethylene (PE) or PE that had been chemically modified to contain superoxide dismutase mimic (SODm, an enzyme with anti-inflammatory properties) were implanted subcutaneously in rats for 3 days. Many neutrophils and inflammatory cells surrounded the PE implants (**A**), but fewer cells surrounded the SODm-PE implants (**B**). **C** illustrates a cross section of a blood vessel near a PE implant; the arrows point to neutrophils in the process of margination and diapedesis. In **D**, a cross section of a blood vessel near a SODm-PE implant, there is no neutrophil activity. Scale bar in **B** = 50 μm (**A** is at the same magnification). Scale bar in **D** = 20 μm (**C** is at the same magnification). Figure reproduced from K. Udipi, R.L. Ornberg, K.B. Thurmond II, S.L. Settle, D. Forster, and D. Riley, "Modification of inflammatory response to implanted biomedical materials in vivo by surface bound superoxide dismutase mimics," *Journal of Biomedical Materials Research*, 51:549–560 (2000), © John Wiley & Sons, Inc., 2000. Reprinted by permission of Wiley-Liss, Inc., a subsidiary of John Wiley & Sons, Inc.

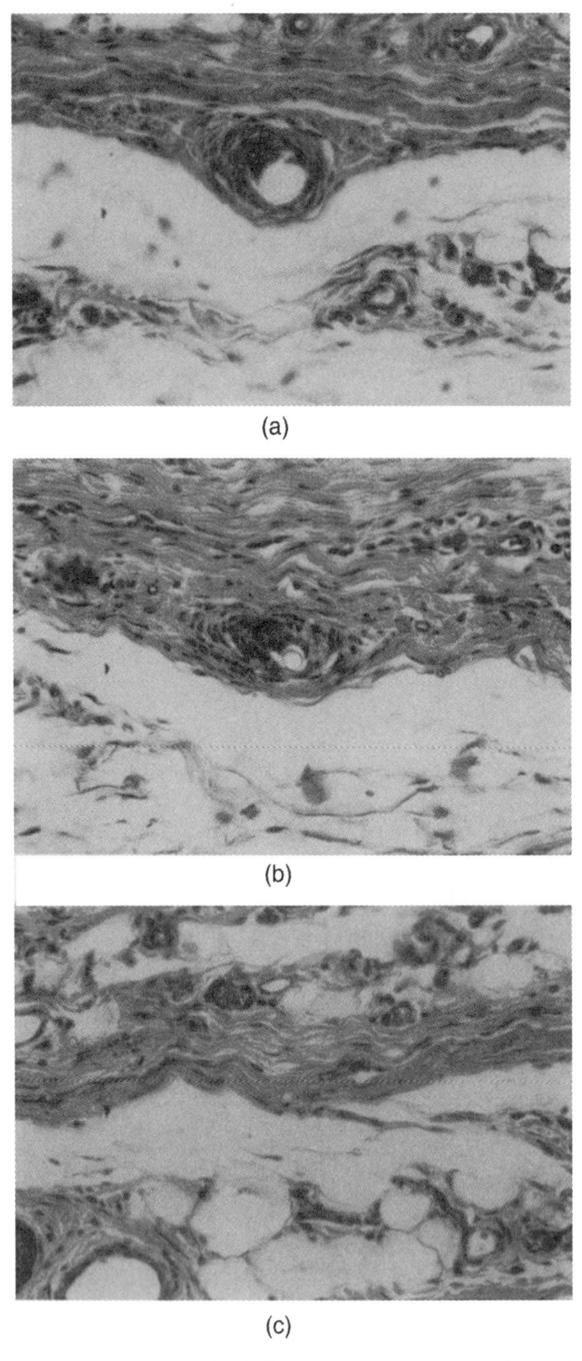

Figure 7.4. The size of implants affects the inflammatory response. Fibers of polypropylene were fabricated with diameters of approximately 26, 12, and 4 μm. **A** through **C** illustrate the histology of the tissues surrounding sample fibers after 5 weeks of subcutaneous implantation in a rat. In each frame, the fiber is in the center of the image and oriented with the long axis perpendicular to the plane of the image ("coming out of the page"); thus the cross section of each fiber in these images is visible as a white circle. In **A** (fiber diameter ~26 μm) and **B** (fiber diameter

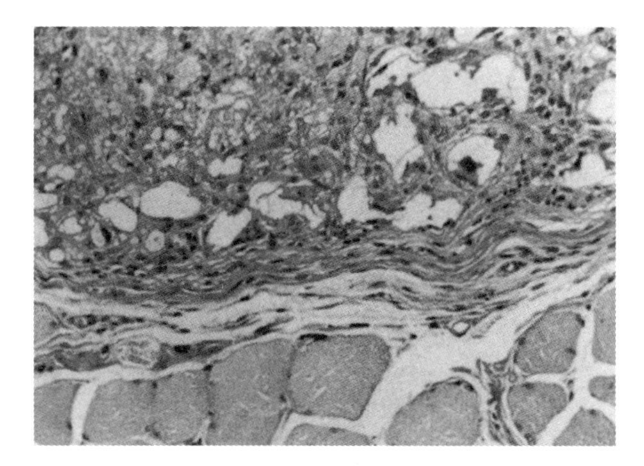

Figure 7.5. Fibrous capsule formation after implantation of a biodegradable material. After 4 weeks of implantation in rat abdominal muscles, histologic examination revealed that an implant containing poly(DL-lactic acid) was surrounded by a fibrous capsule. Unabsorbed particles of the poly (DL-lactic acid) at the bottom of the figure were blanketed by the capsule (wavy fibrillar layer). Above the capsule, large inflammatory cells were infiltrating the smaller cells of the abdominal muscle. Original magnification: ×340. Figure reproduced from E. Solheim, B. Sudmann, G. Bang, and E. Sudmann, "Biocompatibility and effect on osteogenesis of poly(ortho ester) compared to poly (DL-lactic acid)," *Journal of Biomedical Materials Research*, 49:257–263 (2000), © John Wiley & Sons, Inc., 2000. Reprinted by permission of Wiley-Liss, Inc., a subsidiary of John Wiley & Sons, Inc.

main part of what is referred to as "foreign body reaction"; components of the granulation tissue, specifically the new capillaries and fibroblasts, are also part of this response.

With a few exceptions (for example, porous implants in bone), fibrous encapsulation is the end stage of the wound healing process around implants. A fibrous tissue, produced by the host, surrounds the implant materials/devises and isolates them from the surrounding biological environment (Figure 7.5). Fibrous capsule formation depends on the degree of the original trauma, the extent of cell death, and the anatomic location of the injury site. The extent of the foreign body response as well as the composition and thickness of the fibrous tissue depend on the shape and the physical properties (such as topography) of the biomaterial surface. The presence of mechanical forces, as well as the rate of release, accumulation and bioactivity of released chemicals and corrosion products from prosthetic devices, affect the thickness of the fibrous cap-

~12 μm), both a fibrous capsule and inflammatory cell activity surround each fiber. Neither capsule nor inflammatory cell activity is evident around the smallest fiber (diameter ~4 μm) tested (**C**). Figure reproduced from J.E. Sanders, C.E. Stiles, and C.L. Hayes, "Tissue response to single-polymer fibers of varying diameters: Evaluation of fibrous encapsulation and macrophage density," *Journal of Biomedical Materials Research*, 52:231–237 (2000), © John Wiley & Sons, Inc., 2000. Reprinted by permission of Wiley-Liss, Inc., a subsidiary of John Wiley & Sons, Inc.

sule. Thin fibrous capsules are associated with rounded, non-moving, chemically stable implants. Smooth surfaces, however, tend to induce a thicker fibrous capsule than do rough or textured surfaces. Compared to the fibrous layer around straight sections, thick fibrous tissue is formed around sharp corners of implants. While the chemical characteristics of the implant surface modulate earlier events in the wound healing process (specifically, protein deposition, coagulation, interactions of inflammatory and tissue cells with the material surface), many aspects of the end-stages of wound healing, specifically, foreign body reaction and fibrous encapsulation, are at least equally dependent on physical and structural characteristics of the implanted device and material.

Other factors that affect the wound healing outcome around implants include the anatomic site of implantation (especially the regenerative capacity of cells in the surrounding tissues), the adequacy of blood supply at this site, preexisting pathological conditions, and infection. The ideal outcome, of course, is timely resolution of the wound healing process, attainment of a steady state (characterized by cessation of wound healing-related biological changes), and, most important, integration of the implant (without a fibrous capsule) into the surrounding biological milieu.

7.6 COMPLICATIONS RELATED TO WOUND HEALING AROUND IMPLANTS

A number of events may complicate, delay, and even stop the process of wound healing around implant materials and devices. For example, upon contact of blood with implant biomaterials components of the complement cascade may be activated and thus trigger inflammatory leukocyte activation. While short-term (acute) inflammation is a necessary part of the healing process, inflammation that persists for weeks and months post implantation (chronic inflammation) can either delay or prevent resolution of the wound healing at the implantation site. Chronic inflammation could become a severe clinical problem, and may necessitate surgical removal of implanted devices.

Exposure of biomaterials in the wound-healing milieu (an environment rich with reactive chemical and bioactive compounds) may cause metal corrosion and polymer degradation resulting in release of ions and monomers (stabilizers, polymerization initiators and emulsifiers); subsequent chemical activity of these compounds may affect the chemistry and change the topography of the material surface (Figure 7.6). On the other hand, an implanted material/device may be the source of inflammatory stimulation for a number of reasons: for example, because the material fails to become encapsulated and adequately "walled off" the rest of the body, or because the material is leaching pro-inflammatory chemicals, or because the material is breaking up into, and "shedding", small particles. Particulate debris (also referred to as *wear debris*) may, for example, be generated at the tissue/implant interface by dynamic loading, that is, the relative

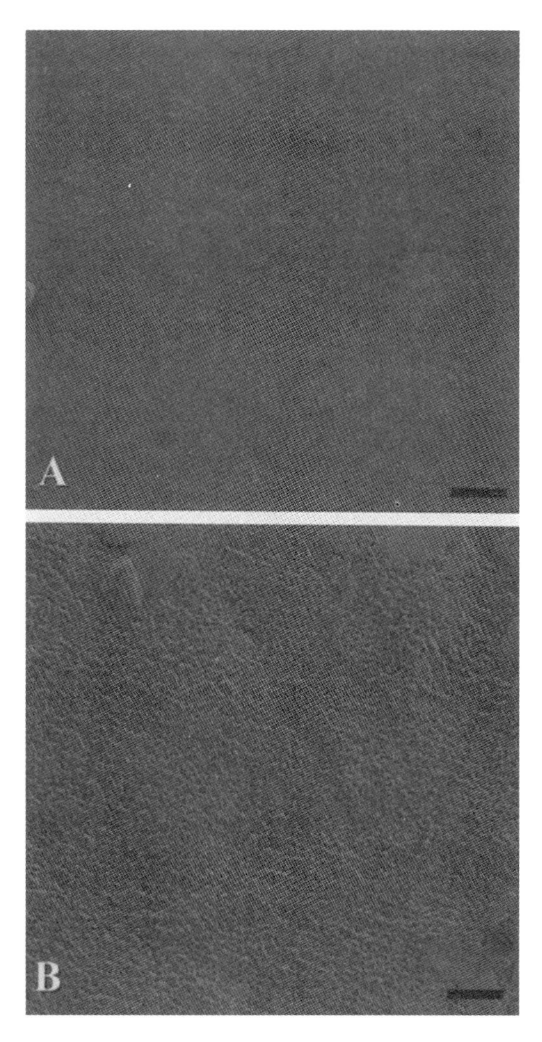

Figure 7.6. Activated inflammatory leukocytes alter the topography of the surfaces of implanted biomaterials. (**A**) Scanning electron micrograph of the surface of a control, nonimplanted poly(etherurethane urea) (PEUU) disk. (**B**) Scanning electron micrograph of the surface of a PEUU disk after 28 days of subcutaneous implantation in a rat. It is believed that the rough texture of the surface developed because of the actions of, and local conditions created by, activated inflammatory leukocytes. Figure reproduced from K. Udipi, R.L. Ornberg, K.B. Thurmond II, S.L. Settle, D. Forster, and D. Riley, "Modification of inflammatory response to implanted biomedical materials in vivo by surface bound superoxide dismutase mimics," *Journal of Biomedical Materials Research*, 51:549–560 (2000), © John Wiley & Sons, Inc., 2000. Reprinted by permission of Wiley-Liss, Inc., a subsidiary of John Wiley & Sons, Inc.

motion and friction (grinding together) of two articulating surfaces (for example, a femoral head and an acetabular cup, the temporomandibular joint, etc.). The type, concentration and biochemical activity of these released compounds may be toxic to cells (cytotoxic), induce inflammation, and impede the timely resolution of the wound healing process locally; most important, these compounds may also trigger inflammatory, allergic, and immune reactions systemically (immunotoxicity). For example, the presence (that is, either type or concentration) of metal ions (such as nickel, chromium, and cobalt) can elicit allergic and/or hypersensitivity responses; such complications have been observed clinically when alloys of metals (such as cobalt-chromium-molybdenum) were used in dental implant applications. The exact mechanism of metal-ion-activation of the immune system is still not completely understood.

For the most part, polymers in current biomedical applications do not elicit significant immunologic responses. Design of biodegradable polymers for temporary residence in the body as well as for tissue engineering applications has focused on developing materials that degrade into byproducts (such as lactic, glycolic, and caproic acids) that are either biocompatible or are already present in the body, and thus can be either excreted or removed via existing metabolic pathways.

Certain metal ions (specifically, nickel) and monomers (for example, benzyl chloride), are chemical carcinogens, which means that through chemical activity these compounds can participate in the transformation of cells from normal to neoplastic phenotype(s). Although there are no long-term epidemiological studies supporting implant-related carcinogenesis, this serious clinical complication cannot be ignored and needs to be determined for each material under consideration for implantation. Development of tumors at or near the surface of an implanted device (referred to as *foreign-body carcinogenesis*) depends on the physical shape/characteristics of the device, as well as on the end-stage foreign body response to the implant. The underlying mechanisms of foreign body carcinogenesis are not fully understood.

Most frequent complications in the wound healing process around implants concern fibrous encapsulation of devices. Large defects due to tissue injury do not heal properly; under these circumstances large amounts of granulation tissue, fibrosis, and scars are formed. Isolation of implant material and devices by fibrous encapsulation is a normal process by which the body copes with the intrusion of implants. Complications, however, may arise when, for example, this relatively thick and impermeable tissue either impairs the mechanical function(s) of implants or, in the case of drug delivery, inhibits release of therapeutic agents. The isolated situation becomes problematic in the case of infection; bacteria and other microorganisms find a haven inside the fibrotic capsule and thrive there untouchable by antibiotics and other medications that either cannot penetrate the fibrous tissue barrier or do so in insufficient amounts to be effective clinically. Under such circumstances, biomedical implants must be removed surgically independent of the performance of the prosthetic devices. In the case of subcutaneous implantation, isolation of the prosthetic device within the fibrous capsule in close proximity to skin may result in extrusion. The implant is literally pushed outside the body. If the implanted biomaterial is biodegradable, the capsule may collapse on resorption of the construct.

7.7 CLOSING REMARKS

Wound healing in tissues like skin is well characterized at the microscopic (that is, cellular/tissue) level, but the underlying molecular level mechanisms (regarding, for example, the signals that trigger quiescent cells at the wound perimeter to proliferate, migrate, and synthesize and deposit new matrix at the wound site) are still poorly understood. Rapid, complete, and perfect (that is, without scarring) repair and restitution of functional, normal tissues after injury as well as control of the various phases (that is, inflammation, repair, remodeling, and fibrous encapsulation) of the wound healing process around biomaterials and implants have yet to be achieved clinically.

Recent advances in cellular and molecular biology have greatly expanded and enhanced current understanding of the biological processes involved in wound healing and tissue regereneration. Of great interest, and promise to reveal critically needed insights into the process, have been studies of fetal wound healing. Embryonic wounds heal rapidly, perfectly (that is, without scarring), and with regeneration of lost tissue. Significant differences between this and the adult model include (but are not limited to) small amounts of transforming growth factor-β, limited angiogenesis, no conversion from fibroblasts to myofibroblasts, and a different mode of wound closure. Further research is undoubtedly needed in this field.

Another noteworthy development has been an ever-growing number of wound care products designed to incorporate the latest understanding of the cellular and molecular level phenomena involved in the dynamic and complex process of wound healing. These innovations aim at alleviating patient suffering, shortening wound healing time periods, and resolving chronic wound healing clinical problems. Such developments have, therefore, the potential of revolutionizing the clinical management of wound healing.

7.8 SUMMARY

- Wound healing, the body's natural response to injury, involves a sequence of highly interdependent processes that overlap in time.
- During the proliferative phase of wound healing, cells from tissues and vasculature at the perimeter of the wound chemotactically migrate into the wound site and proliferate.
- Extracellular matrix components mediate cell-matrix interactions (for example, adhesion) and functions (such as migration and proliferation) and thus play a critical role in the wound healing process.
- A number of growth factors and cytokines (released from activated platelets and leukocytes as well as from cells in tissues adjacent to wounds) modulate important cell functions (such as migration, differentiation, and proliferation) and thus help regulate the wound healing process.

- During the remodeling phase, cells degrade the granulation tissue at the wound site and replace it with tissue that is structurally and composition-wise different than granulation, but more similar to the original, tissue.
- Wound healing around implant devices and materials in adults involves the foreign body response and results in formation of fibrous tissue that encapsulates implants.
- Complications that impede and/or delay the timely resolution of the wound healing process may cause serious clinical problems and necessitate surgical removal of implants.

7.9 BIBLIOGRAPHY/SUGGESTED READING

Anderson, J.M. (1993). Mechanisms of inflammation and infection with implanted devices. *Cardiovasc. Pathol.*, 2(Suppl.):335–415.

Anderson, J.M., Gristina, A.G., Hanson, S.R., Harker, L.A., Johnson, R.J., Merritt, K., Naylor, P.T., and Shoen, F.J. Host reactions to biomaterials and their evaluation. In *Biomaterials Science: An Introduction to Materials in Medicine*, Ratner, B.D., Hoffman, A.S., Schoen, F.J., Lemons, J.E., (eds.), Academic Press, New York, NY (1996), pp. 165–214.

Clark, R.A.F. (ed.). The molecular and cellular biology of wound repair. Plenum Press, New York (1996).

Singer, A.J. and Clark. R.A.F. (1999). Mechanisms of disease: cutaneous wound healing. *New England J. Med.*, 34:738–746.

Yannas, I.V. Tissue and organ regeneration in adults. Springer, New York (2001).

7.10 QUIZ QUESTIONS

1. What are the components of tissues? Give examples of constituents of a “hard” and a “soft” tissue.

2. A plastic, nonporous material (with dimensions $1 \times 1 \times 0.5$ cm^3) is implanted under the skin (1) Describe the short-term (that is, minutes up to 1 day) responses of white blood cells at the site of implantation. (2) Describe the normal, physiological responses of cells in the surrounding connective tissues a few days to a few weeks after surgery. (3) Discuss the pathologic complications that are pertinent and probable for this case.

3. A vascular graft is implanted at the site of a diseased blood vessel. Describe the process of normal wound healing. Discuss potential complications pertinent to this case.

4. An artificial hip prosthesis is implanted in bone. Describe the process of normal wound healing. Discuss potential complications pertinent to this case.

5. Keeping in mind pertinent aspects of various stages of the wound healing process (discussed in Chapters 4–7), explain why tissues that are avascular tend not to heal well.

7.11 STUDY QUESTIONS

1. Polymeric particles (characteristic dimension approximately 1 μm) are produced by the relative motion of a hip prosthesis a few months after implantation.
   Describe the body's response to this occurrence.

2. Specimens of tissue surrounding a hip prosthesis were taken a few months after implantation. Histologic examination revealed a very large number of neutrophils in these specimens.
   What is your assessment of the outcome and/or progress of the wound healing process? What are the implications of such a finding?

3. A vascular graft (synthetic polymer tube) was surgically implanted to replace a segment of a diseased blood vessel.
   Describe the long-term responses of the vascular tissue at the anastomotic sites. How can these events/developments be explained?

4. Choose one extracellular matrix component from those listed in Table 7.2, and look in the recent scientific literature to determine whether, and how, it is being used in tissue engineering research.

5. Choose one growth factor listed in Table 7.1, and look in the scientific literature to learn about its mechanism(s) of action on a specific cell line. What other cell types are affected by this growth factor? What other growth factors have similar effects?

6. Search the scientific literature and identify a paper reporting on the effects of metal ions on the functions of cells *in vitro*. What materials, that are currently used as implants, might release the metal ions that were tested in the reported study? What cell functions were tested? How were these effects quantified? Can the observed *in vitro* results be correlated to processes that occur *in vivo* following implantation of these metals? How might they be different?

7. Collect all the evidence you can in order to determine whether silicone implants cause (or increase the severity of) autoimmune disorders. What do you think? Do they or don't they?

8

Biomaterial Surfaces and the Physiological Environment

8.1 INTRODUCTION

Biomaterials interact with the body through their surfaces. Consequently, the properties of the outermost layers of a material are critically important in determining both biological responses to implants and material responses to the physiological environment. Changes in surface characteristics during exposure to the hostile physiological environment further modify biological responses. A goal of biomaterials research is to design surfaces that elicit desired interfacial behaviors. This chapter describes techniques for characterizing surfaces, approaches for modifying surfaces to control biological responses, and responses of surfaces to wound healing processes.

8.2 SURFACE CHARACTERIZATION METHODS

Surface analytical techniques provide information about the outermost one to ten atomic layers of a material. Characterization of a material's surface properties is needed to relate important surface characteristics to biological responses. Only by thoroughly characterizing implants can biomaterials researchers begin to identify mechanisms by which the body translates atomic and molecular level surface information and uses that information in developing host tissue responses.

Chemical, topographic, mechanical, and electrical properties may all affect how proteins and cells interact with a material. Therefore, comprehensive characterization of a surface requires several pieces of information. Unfortunately, one technique is not capable of providing all the needed information; thorough surface characterization requires the use of multiple analytical methods. The following sections are not exhaustive, but they present techniques commonly used for biomaterial surface characterization.

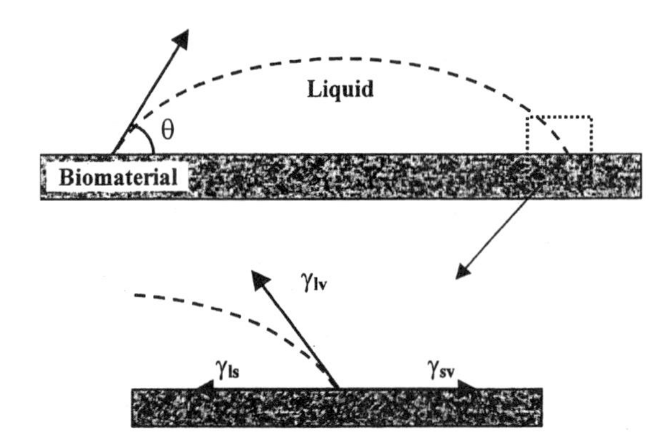

Figure 8.1. Contact angle analysis assesses the ability of a liquid to spread on a surface. The γs represent interfacial surface tensions between the solid and vapor (sv), liquid and solid (ls), and liquid and vapor (lv). The angle of intersection between γ_{lv} and γ_{ls} is the contact angle, θ.

8.2.1 Contact Angle Analysis

Contact angle analysis involves measuring the angle of contact (θ) between a liquid and a surface (Fig. 8.1). When a drop of liquid is placed on a surface, it will spread to reach a force equilibrium, in which the sum of the interfacial tensions in the plane of the surface is zero (Fig. 8.1 and *Eq. 8.1*):

$$\gamma_{sv} - \gamma_{ls} - \gamma_{lv} \cos \theta = 0 \tag{8.1}$$

where γ_{sv} represents the solid-vapor surface tension, γ_{ls} represents the liquid-solid surface tension, and γ_{lv} represents the liquid-vapor surface tension. The contact angle is an inverse measure of the ability of a particular liquid to "wet" the surface. If the liquid is water, a smaller θ indicates a hydrophilic surface, on which water spreads to a greater extent; a larger θ indicates a hydrophobic surface, on which water beads up.

Knowledge of the wettability of a material may have some use, but determination of a material's surface energy (γ_{sv}) better indicates surface properties. Surface energy, defined as the increased free energy per unit area for creating a new surface, is directly proportional to the tendency of molecules to adsorb. Zisman analysis is commonly used to approximate γ_{sv}. Values of θ for a series of liquids are plotted against liquid surface tension (Fig. 8.2). Extrapolation of the fitted line to $\cos \theta = 1$ (where complete spreading occurs) gives the critical surface tension, γ_c.

Although contact angle analysis provides knowledge about the way a surface interacts with gases, liquids, and possibly biomolecules, the information is nonspecific and does not indicate particular chemical or other characteristics of the surface. Its main advantages lie in the relative ease and low cost of the analysis.

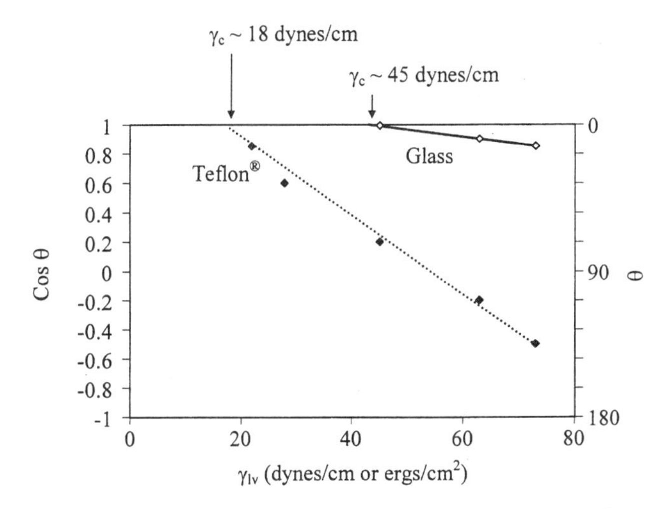

Figure 8.2. Zisman plot for determining the critical surface tension, γ_c, of a material. Example data are shown for both high (Teflon®)- and low (glass)-γ_c surfaces.

8.2.2 X-Ray Photoelectron Spectroscopy (XPS)

XPS, also known as electron spectroscopy for chemical analysis (ESCA), is based on the process of photoemission (Fig. 8.3). On irradiation of a sample with a beam of monochromatic X rays, the X rays penetrate the surface and transfer energy to valence and core level electrons in the sample. If sufficient energy is transferred, electrons will be ejected from the surface. A simple energy balance describes this process:

$$E_b = h\nu - E_k \tag{8.2}$$

where E_b is the binding energy of the emitted electron, $h\nu$ is the energy of the incident X rays, and E_k is the kinetic energy of the ejected electrons. The de-

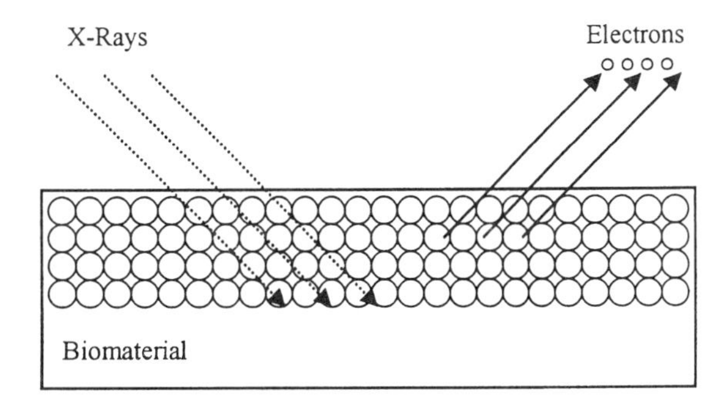

Figure 8.3. Principle of X-ray photoelectron spectroscopy (XPS), in which irradiation of a surface with X rays causes the emission of photoelectrons.

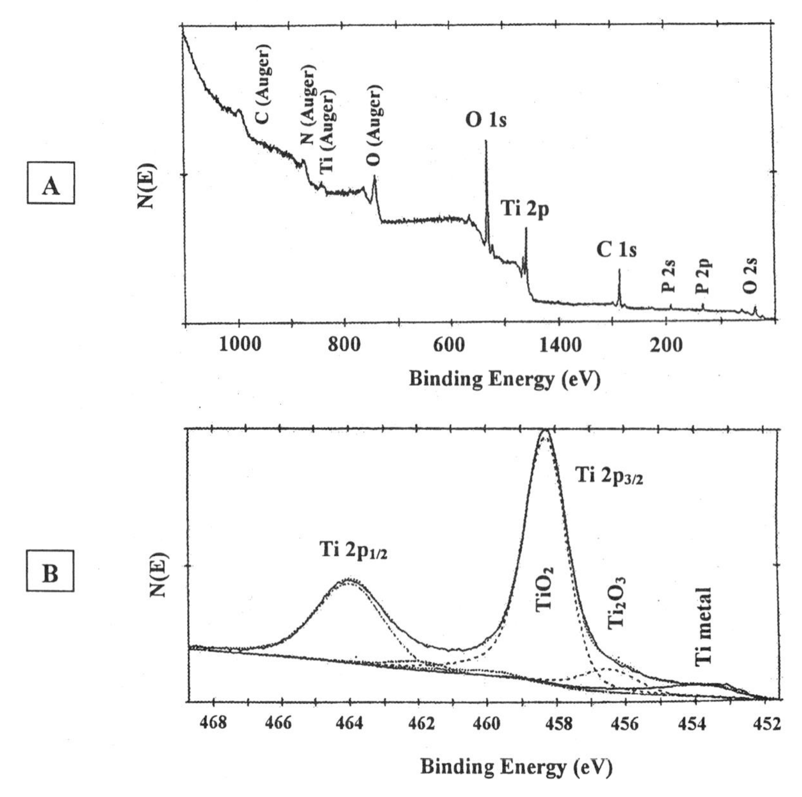

Figure 8.4. XPS spectra for unalloyed titanium. (**A**) Survey spectrum, which primarily shows O, Ti, and C but also minor amounts of P and N. (**B**) High-resolution spectrum of the Ti 2p region, showing primarily Ti^{4+} (TiO_2) but also minor amounts of Ti^{3+} (Ti_2O_3) and Ti^0 (Ti metal). From D.V. Kilpadi et al., *J Biomed Mater Res*, 40:646–659, Copyright © 1998. Adapted by permission of Wiley-Liss, Inc., a division of John Wiley & Sons, Inc.

tectors in XPS instruments measure the number and kinetic energy of emitted photoelectrons. Then, because hν is known, the binding energy can be calculated. An XPS spectrum represents the photoelectron energy distribution (Fig. 8.4A). Because the core electrons of each element have characteristic binding energies, the peaks in the XPS spectra allow identification of all elements, except H and He. With appropriate elemental sensitivity factors, approximate atomic concentrations can be calculated from the relative intensities of the peaks. Additionally, because the electron binding energy is determined by the local chemical environment as well as the type of atom, shifts in the peaks can be used to obtain information about the chemical bonding state of atoms. For example, the oxidation states of species present on the surface of a metallic biomaterial can be differentiated (Fig. 8.4B). Similarly, different carbon-containing species can be distinguished on the surface of polymeric biomaterials (Fig. 8.5).

XPS is conducted without special preparation of samples, but the procedure is carried out in an ultrahigh vacuum environment (typically, 10^{-9} torr). Con-

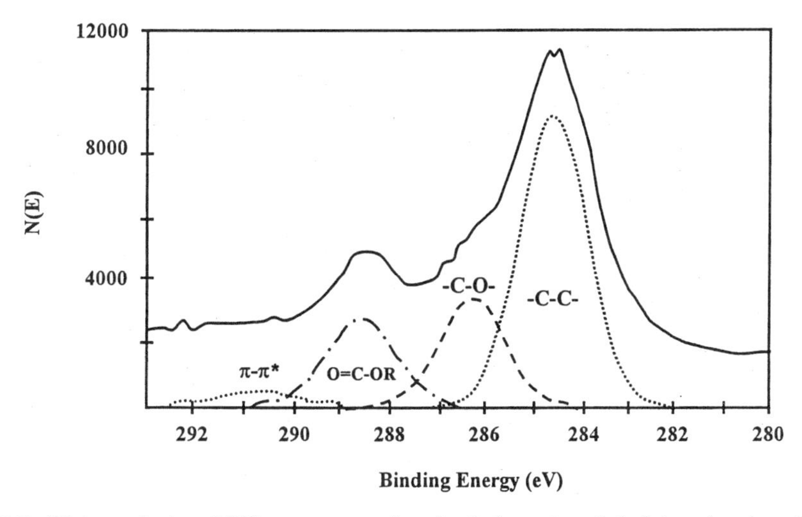

Figure 8.5. High-resolution XPS spectrum of polyethylene terephthalate, showing the contributions of the oxidized carbon species. (From A.J.A Klomp et al., *J Appl Polym Sci* 75:480–494, Copyright © 2000. Adapted by permission of Wiley-Liss, Inc., a division of John Wiley & Sons, Inc.

sequently, biomaterial samples normally must be in a dry state for analysis, which precludes analysis of biological samples, for example, adsorbed proteins, in their native condition. State-of-the-art instruments, however, can use a liquid nitrogen-cooled stage to permit analysis of frozen hydrated samples.

Although X rays can penetrate materials to depths of 1 μm or more, XPS provides information about the outermost 5–75 Å of a material's surface. This is because photoelectrons originating deeper in the sample lose energy in inelastic collisions and/or do not have sufficient energy to be emitted from the sample.

Chemical information as a function of depth into the surface can be obtained in two ways: by sputtering (etching) away surface layers by bombardment with energetic ions before analysis and by changing the angle at which photoelectrons are detected (Fig. 8.6). In the latter, shallow "takeoff" angles reduce the depth of analysis because the photoelectrons cannot escape from deeper atomic layers.

8.2.3 Fourier Transform Infrared (FTIR) Spectroscopy

FTIR spectroscopy is based on the interaction of infrared radiation with a material. Unlike XPS, in which photoelectrons are emitted from the sample, FTIR techniques analyze molecular vibrations induced by the radiation. Imagine atoms as masses joined by bonds with springlike properties (Fig. 8.7). Some atoms have large mass, and others have small mass; some bonds are stiff, and others are easily deformed. When molecules are exposed to infrared light, radiation at frequencies matching the fundamental modes of vibration is absorbed, with the additional restriction that absorption occurs only for those vibrations

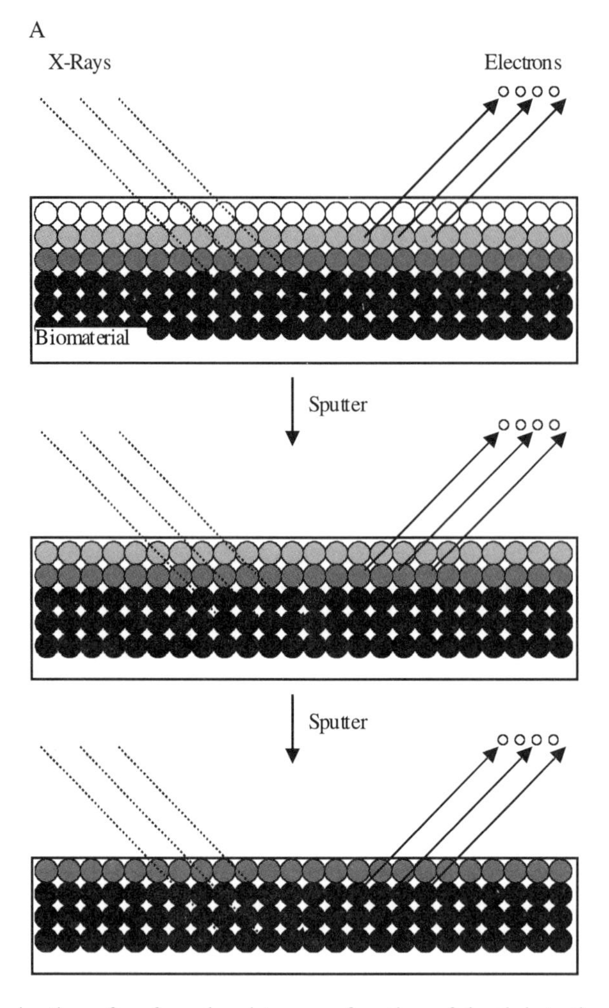

Figure 8.6. Determination of surface chemistry as a function of depth into the surface with XPS. (**A**) Removal of surface layers by sputtering (etching) allows analysis of deeper regions of the material.

that induce oscillating dipoles perpendicular to the surface. IR spectra show peaks (or troughs, depending on how the results are plotted) corresponding to the frequencies at which radiation is absorbed (Fig. 8.8). By using the Fourier transform, absorbance at each frequency can be rapidly determined with a broadband radiation source without sequentially scanning through individual frequencies. Because groups of atoms have unique fundamental modes of vibration, the peaks in a FTIR spectrum represent specific chemical bonds and chemical functional groups; each IR spectrum is a "fingerprint" for that material.

A major attribute of IR analysis is that nearly any type of sample can be studied. Samples can be gas, liquid, or solid. Consequently, FTIR spectrometers

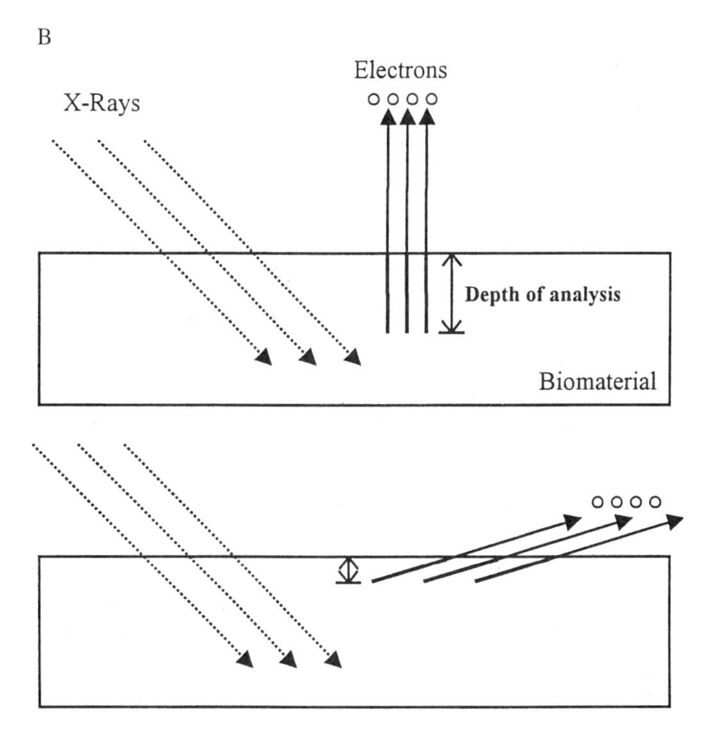

Figure 8.6. (**B**) A smaller angle between the surface and photoelectron detector decreases the depth of analysis.

are available in many configurations. Information obtained by IR spectroscopy, however, typically reflects more than just the outermost atomic layers, generally from 1,000 Å up to 1 μm.

8.2.4 Secondary Ion Mass Spectroscopy (SIMS)

In SIMS, samples are bombarded by monoenergetic ions. The primary ions, frequently oxygen or argon ions, penetrate the surface and transfer energy through a collision cascade, with some particles (secondary particles) gaining sufficient energy to be ejected from the material (Fig. 8.9). The secondary particles can be atoms, molecules, or fragments of the material in charged or neutral states. The secondary ions are analyzed with a mass spectrometer, which allows identification of all elements as well as their isotopes. The spectra of both positive and negative ions are collected to reflect the composition of the material (Fig. 8.10). Interpretation of the spectra, however, is often difficult, because different species can have the same mass numbers (mass-to-charge ratio). For example, both NH_2^+ and O^+ ions have the same position (16) in the positive ion

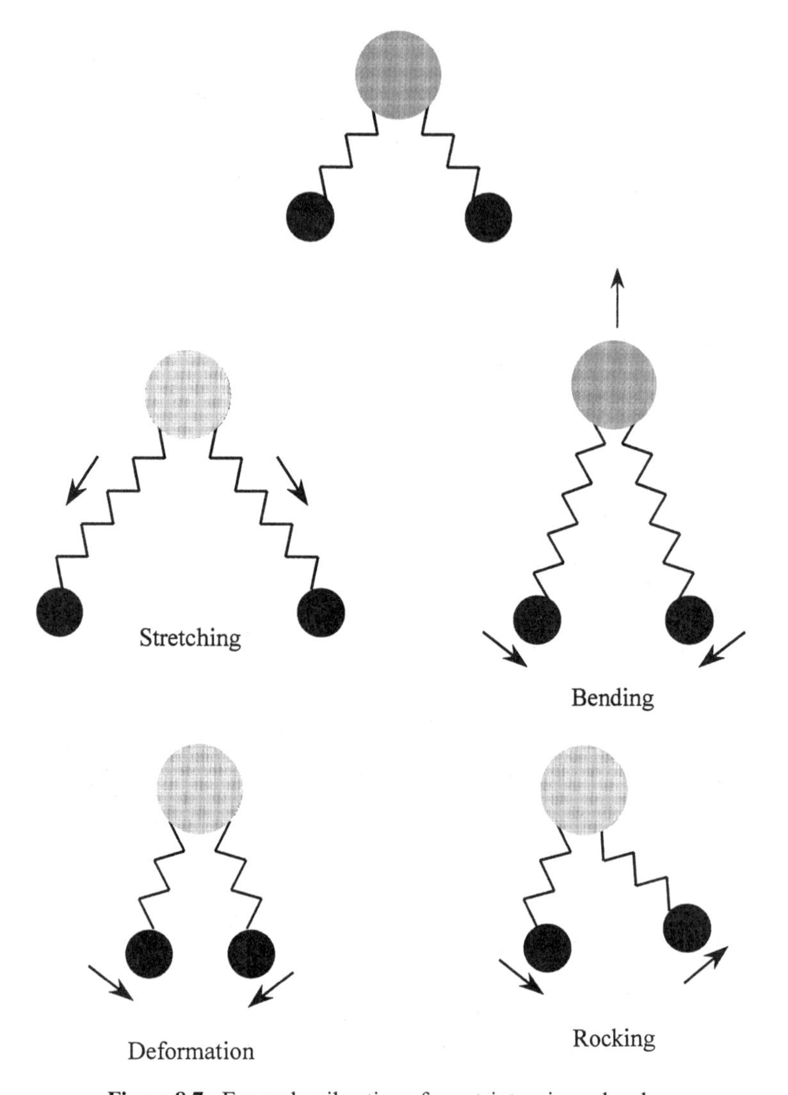

Figure 8.7. Example vibrations for a triatomic molecule.

spectrum. Therefore, reference samples and background knowledge of the samples are needed for meaningful interpretation.

SIMS can be conducted in two modes. In static SIMS, surface sensitivity is ensured by maintaining the ion beam at low intensity. Ideally, each ion impacting the surface will hit areas previously undisturbed. The outermost one to two atomic layers (5–10 Å) can be characterized with static SIMS. With dynamic SIMS, bombardment with high-intensity primary ions causes emission of secondary ions from continuously increasing depth into the material.

As with XPS, SIMS is typically run under ultrahigh vacuum, but a cold stage can be used to examine frozen hydrated samples. Although interpretation of spectra is difficult, SIMS has the benefits of full elemental coverage and great surface sensitivity (down to 10^{-5} atomic %).

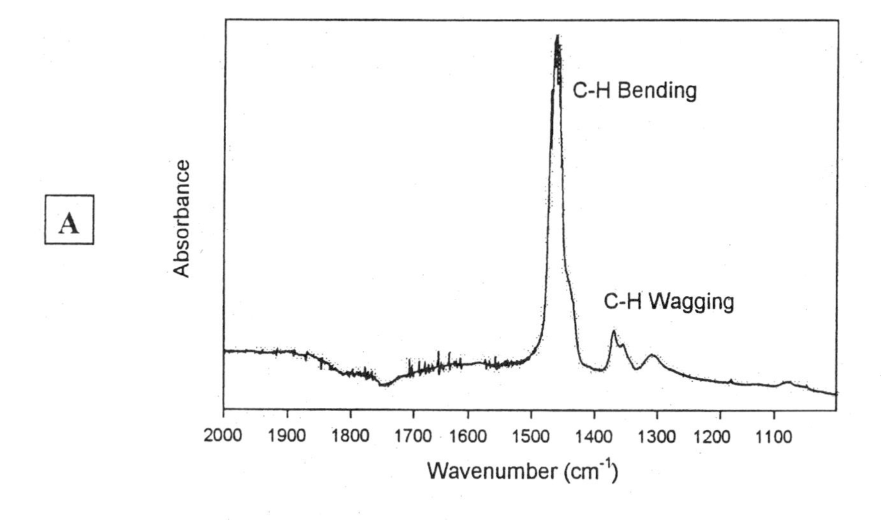

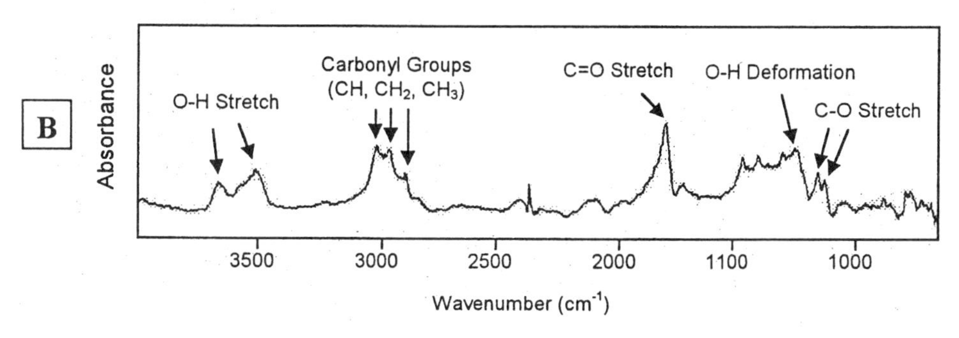

Figure 8.8. FTIR spectra. (**A**) Spectrum for ultrahigh-molecular-weight polyethylene showing absorbance peaks primarily attributed to hydrocarbon vibrations (courtesy of Dr. D. Pienkowski, University of Kentucky). (**B**) Spectrum for poly(lactide-*co*-glycolide) showing absorbance peaks attributed to vibrations of oxygen-containing functional groups in addition to hydrocarbon vibrations (from F.W. Cordewener et al., *J Biomed Mater Res*, 50:59–66, Copyright © 2000. Adapted by permission of Wiley-Liss, Inc., a division of John Wiley & Sons, Inc.).

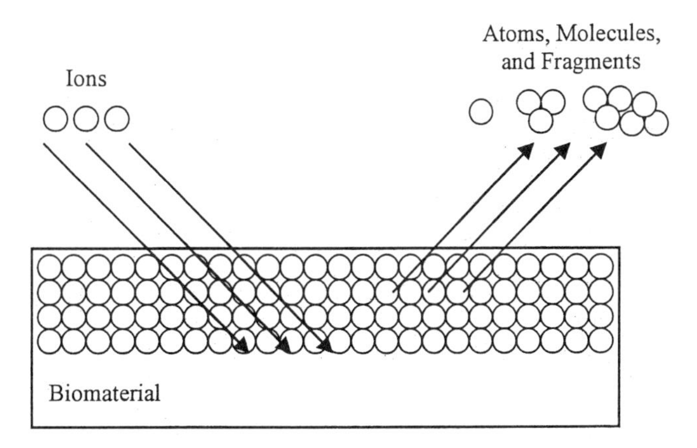

Figure 8.9. Principle of secondary ion mass spectroscopy (SIMS), in which irradiation of a surface with ions causes the emission of atoms, molecules, and fragments.

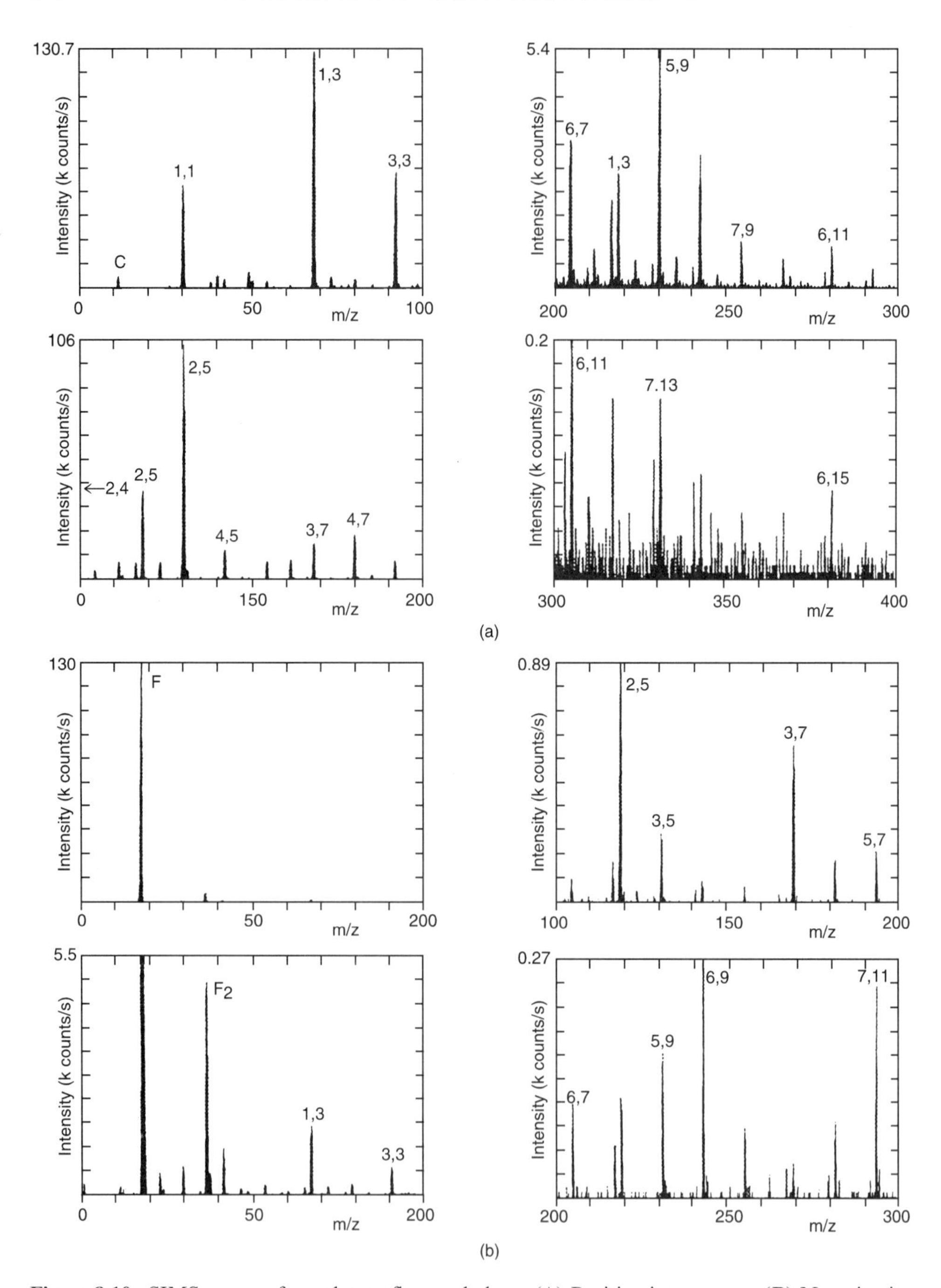

Figure 8.10. SIMS spectra for polytetrafluoroethylene. (**A**) Positive ion spectra. (**B**) Negative ion spectra. The numbers above the peaks indicate the values of x and y in $C_xF_y^+$ and $C_xF_y^-$. From D. Briggs in *Practical Surface Analysis*, D. Briggs and M.P. Seah (eds.), 2nd edition, Vol. 2, pp. 367–423, 1992. © John Wiley & Sons Limited. Reproduced with permission.

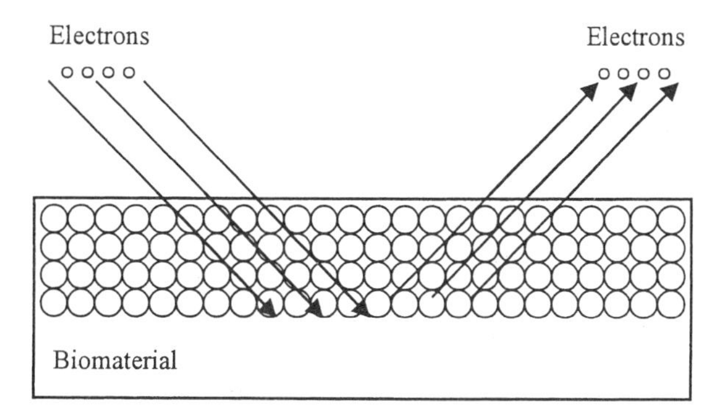

Figure 8.11. Principle of scanning electron microscopy (SEM), in which bombardment of a surface with electrons causes the emission of secondary electrons.

8.2.5 Scanning Electron Microscopy (SEM)

In SEM, a beam of relatively high-energy electrons is scanned across the sample's surface. The primary electrons penetrate the surface and transfer energy to the material in a manner analogous to the way X rays and ions act in XPS and SIMS, respectively. In SEM, the incident electrons transfer sufficient energy for electrons (secondary electrons) to be emitted from the sample (Fig. 8.11). The intensity of the secondary electrons primarily depends on the topography of the surface. By scanning the electron beam across the samples and determining the current generated from secondary electrons, images of the surface are obtained (Fig. 8.12). Thus, in contrast to the methods described previously, which pro-

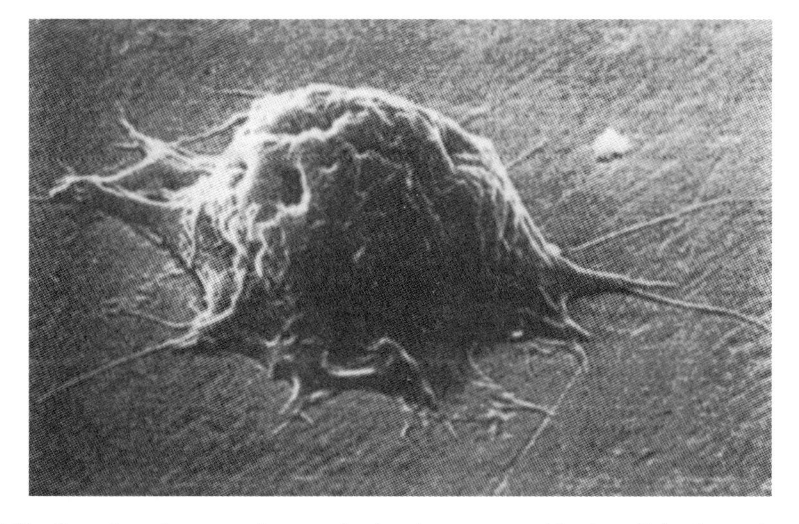

Figure 8.12. Scanning electron micrograph showing an osteoblastic cell in the early stages of spreading on a Ti-6Al-4V surface.

vide surface chemical information, SEM generally gives images reflecting surface topography. Backscattered electrons, which are primary electrons that are elastically scattered back from the sample's surface, however, carry some chemical information, but it is not specific. Run in backscattered electron mode, the information is reflected simply in the form of brighter (higher atomic number) and darker (lower atomic number) regions in the images. More specific chemical information can be obtained with complementary techniques such as energy- or wavelength-dispersive X-ray analysis. The basis of these methods is the emission of X rays following ejection of secondary electrons. The energy or wavelength of the X rays is characteristic of the elements from with they originate.

Because the primary electrons penetrate micrometers into the material and therefore eject secondary electrons from deeper in the material, SEM and the dispersive X-ray analysis techniques are not as surface sensitive as other methods described previously. Also, nonconductive samples, such as polymers and biological materials, must be coated with a conductive film to prevent buildup of negative charge in the sample. Standard SEM is conducted in a high-vacuum environment, which prevents biological samples from being investigated in their native state. Newer instruments, called environmental scanning electron microscopes, allow visualization of at least partially hydrated samples. Nonetheless, SEM is a widely used technique for visually examining surfaces.

8.2.6 Atomic Force Microscopy (AFM)

In AFM, a sharp tip attached to a cantilever is scanned across a surface (Fig. 8.13A). Changes in surface topography that are encountered as the tip moves over the material's surface change the interatomic attractive or repulsive forces between the surface and tip. These forces are sensed by deflection of the cantilever on which the tip is mounted. Two common modes of operation are 1) to vary the tip-surface distance to maintain constant interatomic force and 2) to maintain constant tip-surface distance with variable interatomic force. The height adjustments or changes in interatomic force are recorded and used to construct images of surface topography. To prevent damage as the tip is scanned across the surface, the tip is oscillated perpendicular to the surface at a high frequency, which minimizes lateral forces on the material.

The resolution of AFM images depends, in large part, on the size of the tip (Fig. 8.13B). A tip sharper than the smallest feature to be imaged will generally provide the best resolution. Under the proper conditions, images showing individual atoms can be obtained. Thus a major feature of AFM is the ability to acquire three-dimensional images with angstrom- or nanometer-level resolution. Furthermore, imaging can be conducted without staining, coating, or other preparation and under physiological environmental conditions. Striking images of surfaces, biomolecules, and cells can be obtained (Fig. 8.14).

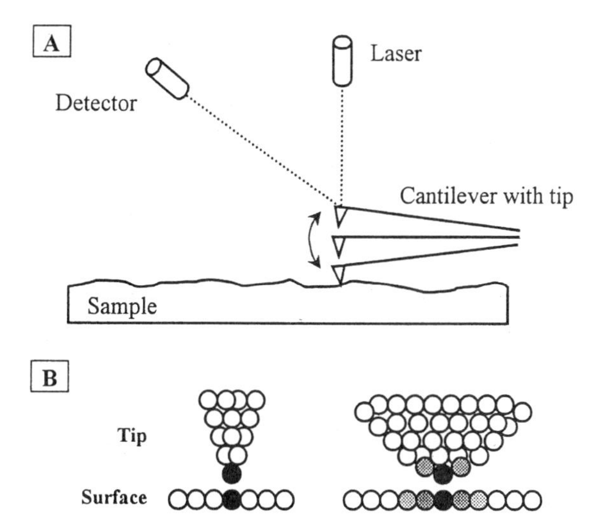

Figure 8.13. Principle of atomic force microscopy (AFM), in which a fine tip is scanned over a surface. (**A**) Schematic illustration of an atomic force microscope. (**B**) Dependence of resolution on the tip size. The dark circles represent atoms with strong interatomic forces, and the shaded circles represent those with weaker interactions. Because of the greater number of atoms interacting with the wider tip, resolution is decreased.

Newer developments in AFM methods enable chemical and mechanical information to be obtained. By attaching specific chemical groups to an AFM tip, the spatial arrangement of functional groups on a surface can be mapped. Also, because AFM is based on interaction between the tip and sample as well as surface topography, local mechanical properties, such as stiffness and friction, can be determined.

8.3 SURFACE RESPONSES TO THE WOUND HEALING PROCESS

When a biomaterial is placed in the body, it is exposed to a complex cascade of events, which is intended to remove or wall off harmful agents and subsequently repair the wound space. The biochemical and cellular mediators of this process, however, can also affect material surface properties. Changes in surface properties can then lead to altered biological responses.

8.3.1 Protein Fouling

Within milliseconds of implantation, biomaterials are covered by adsorbed proteins. Proteins are immediately available in the body fluids, such as blood, that fill the wound site. Additional proteins are secreted by cells associated with wound healing processes. For example, degranulation of platelets during the

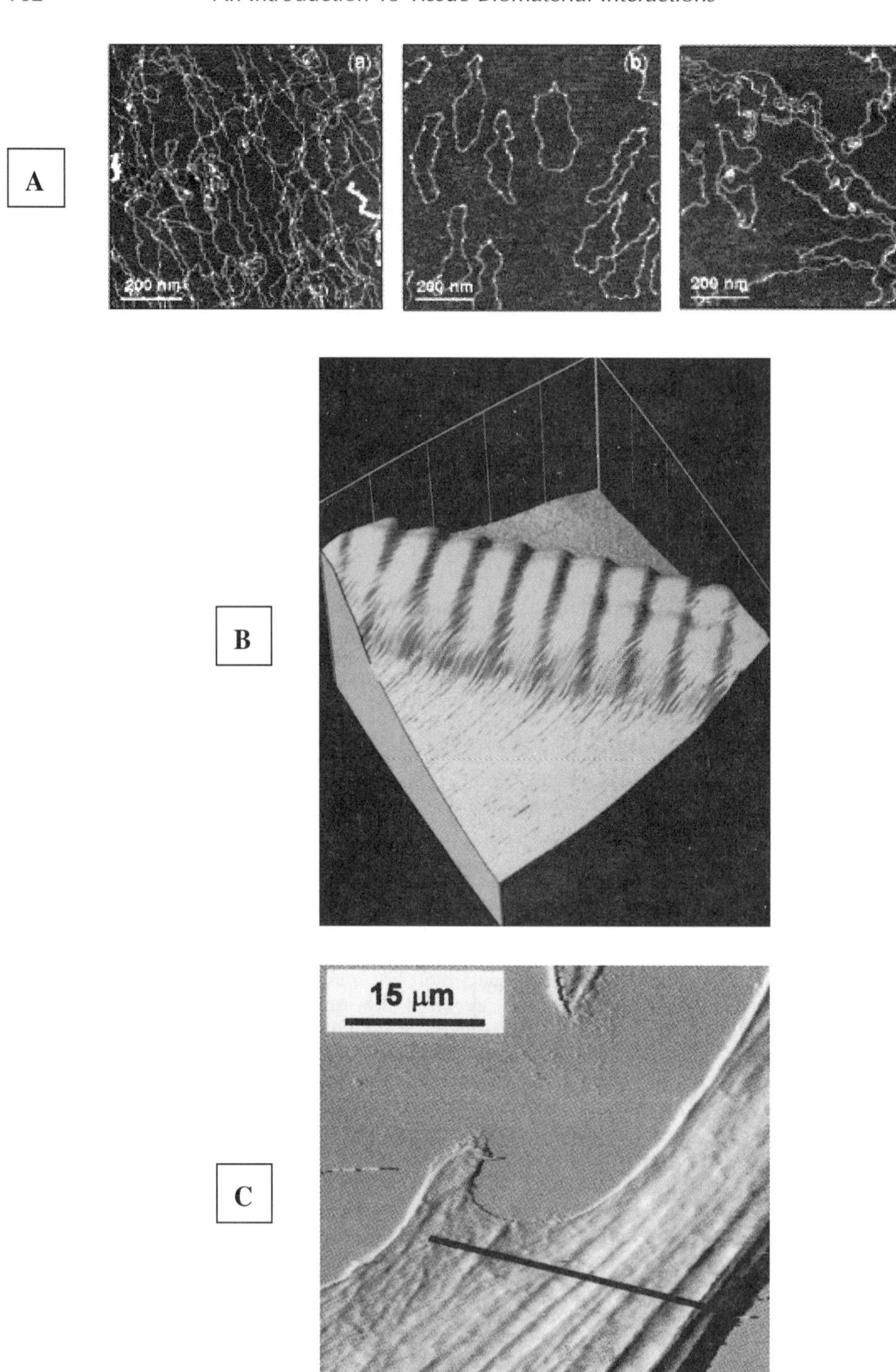
A
(a)
200 nm
(b)
200 nm
(c)
200 nm
B
C
15 μm

release reaction of hemostasis provides numerous chemotactic, bactericidal, and growth-related factors that can interact with the surface. The subsequent arrival of other humoral factors, such as antibodies, and leukocytes is an additional source of biomolecules for adsorption on the biomaterial surface. The particular proteins and the sequence in which they adsorb depend on the specific biomaterial, as described in Chapter 3. An important fact, however, is that the implant's surface is almost immediately converted from a "bare" biomaterial to one fouled with biomolecules. Consequently, the adsorbed proteins mediate biological responses.

8.3.2 Degradation and Dissolution

Although selected for their ability to resist degradation in the physiological environment, biomaterials are not immune from the damaging influences of the body. Its warm neutral saline environment with dissolved oxygen, cells, enzymes, etc. harms metals, polymers, and ceramics.

Corrosion is the deterioration and removal of a metal by chemical attack. Metal atoms are ionized and go into solution, where they may react with oxygen or form complexes with biomolecules. Even the most corrosion-resistant metals can be affected by implantation in the body. Variations in the microstructure of an implant's surface influence removal of ions. For example, the higher-energy grain boundaries are more reactive and therefore are more readily attacked than the grain interior. Certain phases within an alloy also may be more susceptible to attack. Adsorbed proteins can act as barriers to diffusion and alter the electrochemical corrosion reactions, for example, by creating oxygen-deficient regions on the surface. Changes in microenvironmental conditions resulting from inflammatory and wound healing processes also affect surface properties. For example, decreased pH at the tissue-implant interface can alter corrosion processes, and inflammatory species such as peroxide can change the nature of the passivating oxide film. The oxide film can grow to several times its original thickness, and S, Ca, and P can be incorporated into the surface layer. The surface reactivity of metals can be further aggravated by both static and dynamic mechanical loading.

Polymeric biomaterials are subject to degradation by several mechanisms. Species such as the superoxide anion ($O_2^{\bullet}$), hydrogen peroxide (H_2O_2), and hypochlorite, which are produced by neutrophils, macrophages, and foreign body

◀

Figure 8.14. AFM images of (**A**) DNA molecules (from D. Dario Anselmetti et al., *Single Mol* 1:53–58, Copyright © 2000. Reprinted by permission of Wiley-VCH), (**B**) a collagen fibril, where the dark regions correspond to the bands observed by transmission electron microscopy (from M. Raspanti et al., *Microsc Res Tech*, 35:87–93, Copyright © 1996. Reprinted by permission of Wiley-Liss, Inc., a division of John Wiley & Sons, Inc.), and (**C**) a cellular lamellipodium (from G.R. Bushell et al., *Cytometry*, 36:254–64, Copyright © 1999. Reprinted by permission of Wiley-Liss, Inc., a division of John Wiley & Sons, Inc.).

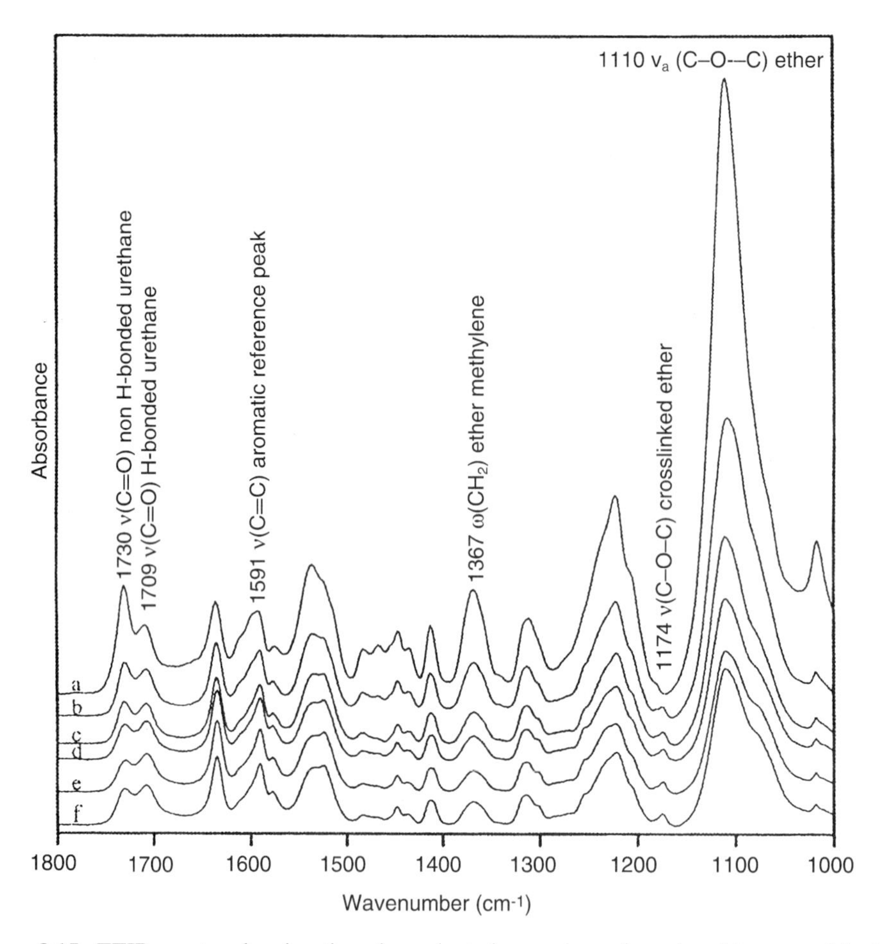

Figure 8.15. FTIR spectra showing time-dependent changes in surface chemistry caused by incubation of a polyetherurethane urea in a solution of hydrogen peroxide to oxidize the surface. (**a**) Untreated, (**b**) 3 days, (**c**) 6 days, (**d**) 9 days, (**e**) 12 days, and (**f**) 15 days of treatment. From M.A. Schubert et al., *J Biomed Mater Res*, 34:493–505, Copyright © 1997. Reprinted by permission of Wiley-Liss, Inc., a division of John Wiley & Sons, Inc.

giant cells, may cause oxidation and subsequent surface chemical changes in polymers (Fig. 8.15). Surface cracking and pitting can also occur, especially when the polymer is mechanically loaded (Fig. 8.16). Hydrolysis can be catalyzed by physiological ions, such as PO_4^{3-}, or by enzymes secreted during the wound healing process. As with oxidation, hydrolysis can result in cleavage of macromolecular chains, resulting in exposure of different functional groups on the surface of the material and/or alteration of surface texture.

Although ceramics are not susceptible to corrosion or oxidative/hydrolytic degradation, their surface properties can also be altered under physiological conditions. Certain calcium phosphate ceramics are disposed to dissolution. Tricalcium phosphate in particular is readily soluble. Even for nearly "inert"

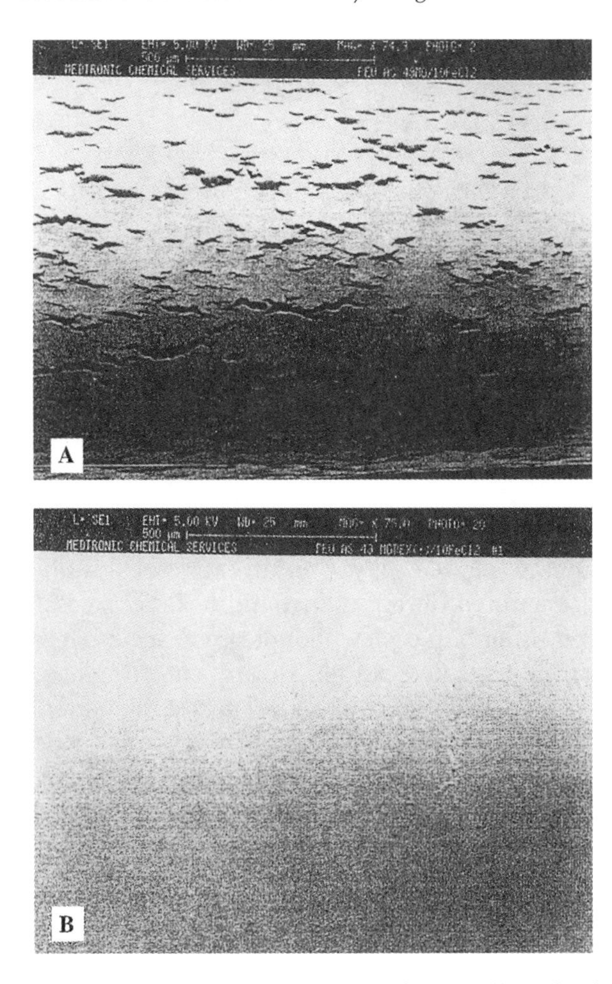

Figure 8.16. Scanning electron micrographs showing surface cracking of polyetherurethane that was cultured with (**A**) human macrophages/monocytes and (**B**) human macrophages/monocytes treated to prevent their production of H_2O_2. From J. Casas et al., *J Biomed Mater Res* 46:475–484, Copyright © 1999. Reprinted by permission of Wiley-Liss, Inc., a division of John Wiley & Sons, Inc.

ceramics, such as alumina, small amounts of ions can be leached from the surface. Removal, especially selective leaching, of material from the implant can alter surface properties, and the ions can subsequently affect biological responses.

8.3.3 Calcification

Deposition of calcium-containing minerals on the surface of a biomaterial can occur after implantation. Certain proteins possess anionic sequences that bind calcium ions, and some are believed to participate in mineralization of the col-

TABLE 8.1. Calcification of Polyurethane Alters the Surface Chemistry Measured by XPS

Polyurethane	Incubation Period (days)	C%	N%	O%	Si%	Ca%	P%
Chronoflex	0	71.3	1.1	22.6	5.0	—	—
	15	71.8	2.8	21.3	3.7	0.1	0.2
	30	67.3	3.1	23.6	4.0	1.1	0.8
	60	57.3	2.5	29.5	2.1	4.3	3.2
Corethane 80A	0	72.8	1.6	25.2	0.2	—	—
	15	68.4	2.3	25.8	1.9	0.9	0.7
	30	69.6	3.3	25.5	0.3	0.9	0.5
	60	68.1	4.3	24.3	1.0	1.1	0.9

From M. Yang et al., *J Biomed Mater Res*, 48:648–659, Copyright © 1999. Adapted by permission of Wiley-Liss, Inc., a division of John Wiley & Sons, Inc.

lagenous matrix of bone. Therefore, adsorption of these calcium-binding proteins on a biomaterial surface may nucleate mineral, leading to calcification. Remnants of cells ruptured during the wound healing process may also play a role in calcification. Membrane phospholipids, intracellular organelles, and enzymes can lead to mineral nucleation. Subsequent growth of mineral crystals causes not only significant chemical changes in the biomaterial surface (Table 8.1) but mechanical changes as well. For example, the increased stiffness caused by calcification can lead to tearing of polymeric materials, such as might be used in synthetic heart valves.

8.4 ENGINEERING BIOMATERIAL SURFACES

"Biomaterial surface engineering" offers the ability to modify material and biological responses through changes in surface properties while maintaining the bulk properties of the implant. Only surface modifications are needed, because biological tissues interact with mainly the outermost atomic layers of a biomaterial. Biomaterial surface engineering approaches can be classified according to the surface properties being altered, e.g., physicochemical, morphologic, or biological modifications.

8.4.1 Morphologic Modifications

Alterations in surface morphology and roughness have been used to influence cell and tissue responses to implants. Coatings containing pores were developed with the objective of encouraging tissue ingrowth, which would increase fixation of implants because of mechanical interlocking. Porous coatings have been studied extensively for use on orthopedic implants, such as hip and knee replacements, and many implants with porous coatings are commercially available (Fig. 8.17A). Other applications include the cuffs of heart valves, where

Figure 8.17. Morphologic modifications of surfaces. (**A**) Scanning electron micrograph of a porous-coated Co–Cr alloy (from R.M. Pilliar, *J Biomed Mater Res*, 21(A1):1–33, Copyright © 1987. Reprinted by permission of Wiley-Liss, Inc., a division of John Wiley & Sons, Inc.). (**B**) Scanning electron micrograph showing cells aligned parallel to 5-μm grooves on a silicone surface (from E.T. den Braber et al., *J Biomed Mater Res*, 29:511–518, Copyright © 1995. Reprinted by permission of Wiley-Liss, Inc., a division of John Wiley & Sons, Inc.).

they must be anchored to the heart muscle, and regions of percutaneous implants where they penetrate the skin. Grooved surfaces can induce "contact guidance," the phenomenon whereby cell orientation and the direction of cell movement are affected by the morphology of the substrate (Fig. 8.17B). This principle has been investigated for preventing epithelial downgrowth on dental implants and directing bone formation along particular regions of an implant.

8.4.2 Physicochemical Modifications

Surface energy, surface charge, and surface composition are among the physicochemical characteristics that have been altered with the aim of changing both

TABLE 8.2. Plasma Treatment of Polyvinyl Alcohol With Acetone-O_2 and Ammonia Changes the Surface Chemistry Measured by XPS

Plasma Treatment	C%	O%	N%	O/C	N/C
Untreated	77.9	19.5	2.6	0.26	0.03
Acetone-O_2	72.4	23.9	3.7	0.33	0.05
NH_3	69.5	26.3	4.2	0.38	0.06

From R. Latkany et al., *J Biomed Mater Res*, 36:29–37, Copyright © 1997. Adapted by permission of Wiley-Liss, Inc., a division of John Wiley & Sons, Inc.

material and biological responses. Glow discharge, a process in which surfaces are exposed to ionized inert gas (a plasma), has been used to increase surface free energy of many metals and polymers. The greater reactivity of surfaces with higher surface energy generally leads to increased tissue adhesion. Because energetic species in the plasma can break bonds, the surface of polymeric materials can also be cross-linked, which decreases surface permeability and increases surface hardness. Plasma treatment with reactive gases, such as water, oxygen, or ammonia, can create new functional groups on polymer surfaces (Tables 8.2 and 8.3). In addition to changing the way proteins and cells interact with the biomaterial, surface mechanical properties may be altered.

Ion implantation has been used to modify both material and biological responses. As high-energy ions impact a surface, a collision cascade is initiated, as previously described for SIMS. The ions not only interact with the surface to cause chemical changes, but the energetic species can induce structural alterations as well. Implantation of N and C ions into metallic biomaterials improves the corrosion and wear resistance of the material. Silver ions have been implanted into polymeric materials to provide surface antimicrobial activity for catheters.

More dramatic changes in surface chemistry can be obtained by grafting macromolecules onto the biomaterial. For example, attachment of poly(ethylene oxide) to surfaces has been studied extensively. Although results depend on molecular surface density and chain length, protein adsorption and, subsequently, cell adhesion can be significantly reduced. The performance of biomaterials used in blood-contacting applications might be improved with this approach.

TABLE 8.3. Plasma Treatment of Polyvinyl Alcohol With Acetone-O_2 and Ammonia Results in Larger Carbonyl (C=O) and Carboxylic (COO) Components in the C 1s XPS Spectrum

Plasma Treatment	%CH_x	%C–O or C–N	%C=O or O–C–N	%COO or O=C–N
Untreated	68.8	26.3	5.9	—
Acetone-O_2	63.2	22.9	10.4	3.5
NH_3	57.3	30.4	9.9	2.4

From R. Latkany et al., *J Biomed Mater Res*, 36:29–37, Copyright © 1997. Adapted by permission of Wiley-Liss, Inc., a division of John Wiley & Sons, Inc.

Considering the role of electrostatic interactions in many biological events, charged surfaces can also modify protein and cell behavior at interfaces. Negatively charged surfaces can be created by deposition of acidic or sulfonate-containing functional groups with grafting techniques, and amino-containing functional groups can be used to produce positively charged surfaces. Surfaces with negative charge tend to delay thrombogenesis, and those with positive charge accelerate it.

Although technically not surface modifications, coatings can be used to provide surface properties drastically different from those of the unmodified material. In orthopedics, for example, calcium phosphate ceramic coatings on metallic implants have been extensively investigated because of their chemical similarity to bone mineral.

8.4.3 Biological Modifications

Biological modification of surfaces uses understanding of the cell and molecular biology of cellular function and differentiation. Much has been learned about the mechanisms by which cells adhere to substrates, and major advances have been made in understanding the roles of biomolecules in regulating cell differentiation. The goal of biological surface modification is to control cell and tissue responses to an implant by immobilizing biomolecules on biomaterials.

Since identification of the Arg-Gly-Asp (RGD) sequence as mediating adhesion of cells to several plasma and extracellular matrix proteins, including fibronectin (see Chapter 2), RGD-containing peptides have been deposited on surfaces to promote cell attachment (Fig. 8.18). Cell surface receptors in the integrin superfamily (see Chapter 7) recognize the RGD sequence and mediate adhesion. Selectivity for particular integrins is being sought by using peptides of different length (from 4 to 15 peptides) and of different conformation (from linear to cyclic peptides).

Heparin/heparan sulfate-binding peptides have also been used to enhance cell adhesion. Because heparan sulfate proteoglycans can be present on cell surfaces, interaction between the negatively charged proteoglycans and positively charged peptides can mediate cell attachment. These peptides can be used alone or in combination with RGD peptides to further promote cell-biomaterial interaction.

Another approach to biological surface modification uses whole biomolecules. Whereas depositing small peptides gives the surface a specific characteristic such as cell binding to RGD peptides, immobilizing proteins can provide many functions because of the various domains within the molecule (see Chapter 2). Intact adhesive proteins, such as fibronectin and laminin, have been immobilized on biomaterials. Attachment of growth factors (see Chapter 7), which can be produced by recombinant DNA techniques, to surfaces (Fig. 8.19) has the potential to give implants the ability to induce cell growth, activity, and/or differentia-

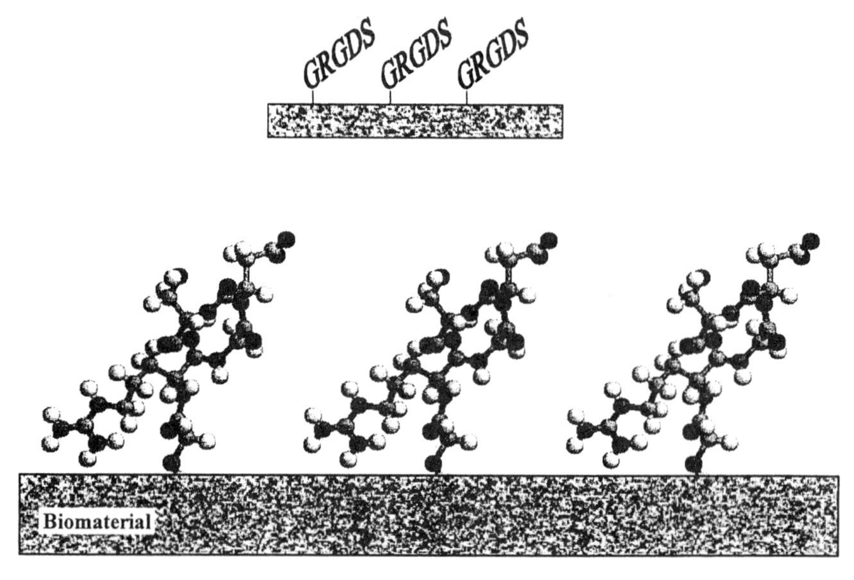

Figure 8.18. Representation of an RGD-containing peptide (GRGDS) immobilized on a surface to enhance cell attachment by binding cell surface receptors of the integrin superfamily.

tion. Growth factors, such as epidermal growth factor, insulin-like growth factor I, and bone morphogenetic protein 4, have been immobilized on biomaterial surfaces to induce specific cellular responses that cannot be obtained with only adhesion-promoting molecules.

Biological surface modification also can be used to produce surfaces resistant to protein and cell adhesion. Phosphorylcholine has been grafted onto or incorporated into biomaterial surfaces to mimic phospholipid head groups of the cell surface (Fig. 8.20). These surfaces have potential for use in blood-contacting applications.

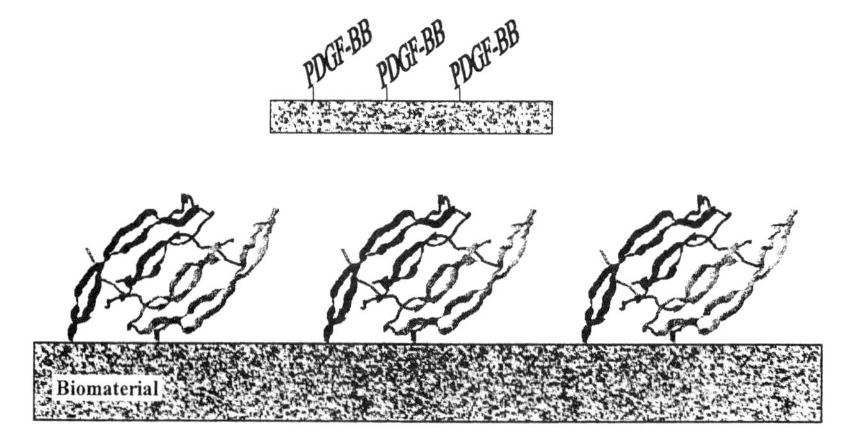

Figure 8.19. Representation of dimeric platelet-derived growth factor-BB immobilized on a surface to induce specific cell responses after binding to cell surface receptors.

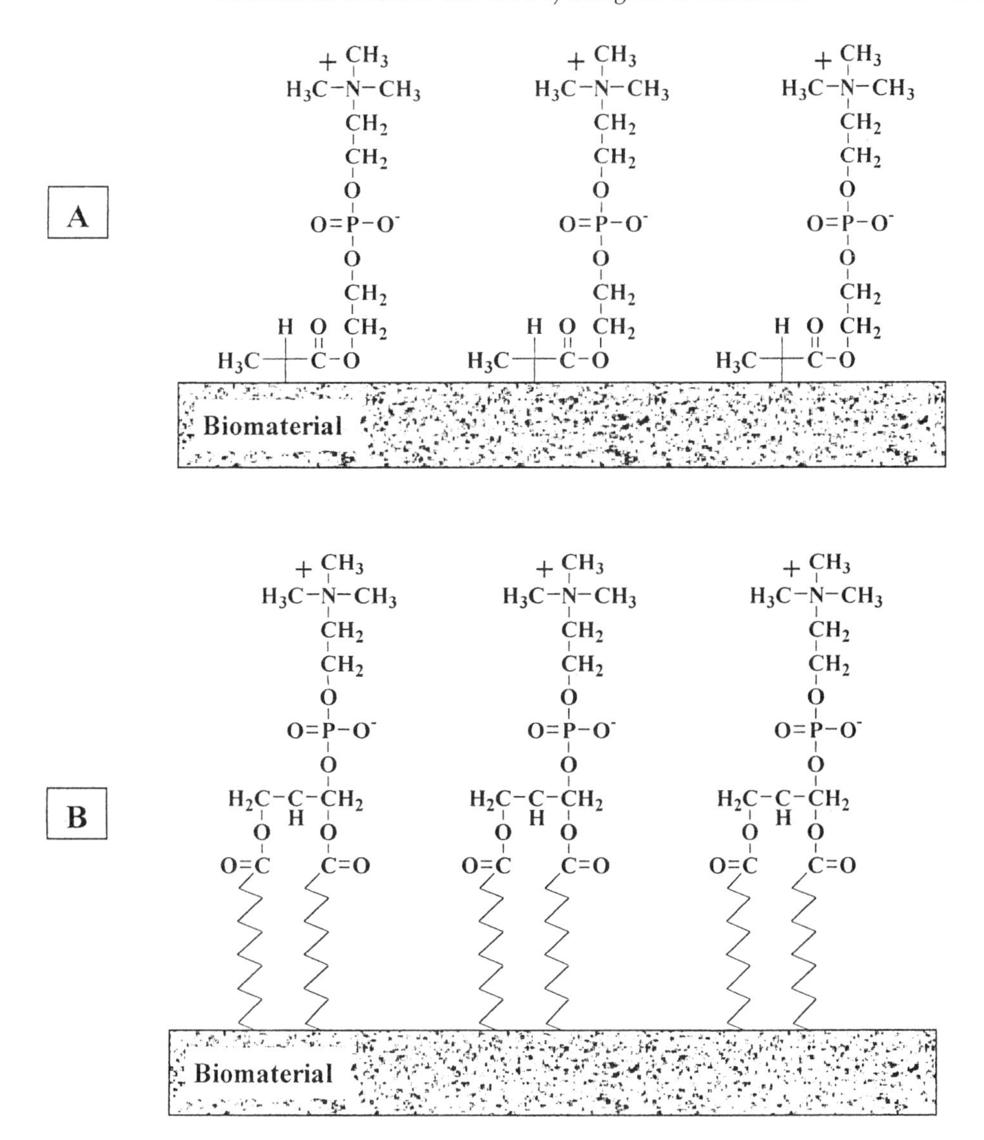

Figure 8.20. Representation of (**A**) phosphorylcholine molecules similar to the polar head group of phospholipids and (**B**) phospholipid-like molecules immobilized on a surface to provide resistance to protein adsorption and cell adhesion by mimicking the cell membrane.

8.5 SUMMARY

- Thorough characterization of a biomaterial surface requires multiple complementary techniques.
- Modern surface characterization methods can provide information about the outermost one to ten atomic layers (<50 Å) of a biomaterial's surface.
- Contact angle analysis, XPS, FTIR, and SIMS are commonly used to obtain surface chemical information.

- SEM and AFM provide topographic information about surfaces.
- Continued developments have expanded the capabilities of surface analytical techniques to provide chemical maps and surface mechanical properties.
- Analysis of biomaterial surfaces after simulated and actual exposure to physiological environments has demonstrated significant alterations in chemistry and topography.
- Biomaterial surface engineering techniques are being used to modify the morphologic, physicochemical, and biological properties of biomaterials in an attempt to obtain desirable tissue-implant interactions.

8.6 BIBLIOGRAPHY/SUGGESTED READING

Kasemo, B. and Lausmaa, J. (1986). Surface science aspects on inorganic biomaterials, *CRC Crit. Rev. Biocomp.*, 2:335–380.

8.7 QUIZ QUESTIONS

1. What is the basis of contact angle analysis? What information does this provide?
2. What is the common fundamental principle of XPS and SIMS? What type of information is provided by these methods?
3. How does FTIR spectroscopy differ from XPS and SIMS?
4. What type(s) of information is provided by AFM?
5. What is the basis of scanning electron microscopy?
6. By what mechanisms can biomaterials degrade in a physiological environment?
7. List surface properties that can be changed to alter events at the tissue-implant interface.

8.8 STUDY QUESTIONS/DISCOVERY ACTIVITIES

1. Discuss the advantages and disadvantages of contact angle analysis.
2. If you were given a novel biomaterial, what method(s) would you use to characterize it? Why?
3. As a biomedical engineer for a large manufacturer of orthopedic implants, you have been asked to modify a biomaterial surface to enhance integration with bone. How would you approach this task?
4. Conduct a literature search on the topic of biomaterial surface characterization. What techniques were commonly used? What materials were investigated? Select a recent publication. Summarize the findings. What was the purpose of determining the surface properties?

9

Biocompatibility

9.1 INTRODUCTION

Consideration of biomaterials and devices for short- or long-term applications that bring them in contact with body fluids, tissues and organs is not complete without evaluation of their biocompatibility. Such testing must examine the performance a biomaterial under conditions that simulate those of the biological milieu as well as the end-use application and duration of exposure to these conditions.

Biocompatibility of devices (and, by necessity, of their constituent materials) must be established (that is, tested and documented by manufacturers) and approved by appropriate regulatory agencies [for example, the Food and Drug Administration (FDA) in the United States] before any biomedical device can be marketed and used clinically. Such practice is in compliance with laws (for example, the *Medical Device Amendment of the U.S. Food and Drug Act* of 1976 and the *Safe Medical Devices Act* of 1990) that require proof of safe and efficacious performance of medical devices under intended-use conditions. Other countries have similar regulations and laws that must be considered when a product is to be marketed internationally.

Biocompatibility studies may be conducted in laboratories at universities and in industry according to established procedures and standards. Such practices, however, require appropriate facilities, adequate resources, and trained personnel as well as pertinent planning, maintenance, and budgets that may be outside the range of scientific priorities and financial means of small and medium-size research groups and companies. The alternative in such cases is to pay fees and have these tests performed by reputable commercial testing laboratories. It should be noted that *all* tests must be conducted according to carefully constructed protocols that include appropriate controls. Without pertinent controls, no conclusions whatsoever can be drawn; such studies are incomplete as well as a waste of valuable time and resources.

9.1.1 Differences between Biological and Synthetic Materials

The characteristic properties of biological and engineered or synthetic materials are distinctly different (Table 9.1). For example, whereas tissues contain cells,

TABLE 9.1. Characteristics of biological and synthetic materials

Biological Materials	Synthetic Materials
Cellular	Acellular
Hydrous	Anhydrous
Anisotropic	Isotropic
Heterogeneous	Homogeneous
Viscoelastic	Elastic
Capable of Self-repair/Alive	Inanimate

metals, ceramics, and polymers do not. Whereas tissues have the ability to partially or completely repair themselves, metals, ceramics, and polymers do not. The differences between tissues and their replacements can be illustrated with the examples shown in Table 9.2. Consider the wall of a blood vessel. The lumen of every blood vessel is lined with endothelial cells. The major subendothelial structural components include smooth muscle cells, collagen, and elastin. The relative amounts of these constituents and the orientation of fibers depend on the specific location within the vascular tissue as well as the type (either artery or vein) and size of the blood vessel. To replace this complex tissue, polymeric tubes made either of polytetrafluoroethylene or poly(ethylene terephthalate) are commonly used as synthetic grafts. The distinct physical, mechanical, and chemical properties between biological and synthetic materials give rise to several considerations when evaluating materials for implantation.

9.2 TESTS PREREQUISITE TO EVALUATION OF BIOCOMPATIBILITY

In vitro characterization of materials and of the functional performance of device prototypes are prerequisites that must be met successfully prior to evaluation of their biocompatibility. Characterization of physicochemical (such as

TABLE 9.2. Tissues and some biomaterials that have been used to replace them

Tissue	Synthetic Material Replacement
Blood vessel	Polytetrafluoroethylene Poly(ethylene terephthalate)
Intraocular lens	Polymethylmethacrylate
Hip	Ti-6Al-4V Co-Cr-Mo
Tooth	Amalgam (filling) Ti (implant)

surface and bulk) and other pertinent (for example, mechanical, electrical, transport, and, if applicable, biodegradation) properties must be performed on raw materials. These data must be compared with results at the end of the manufacturing, sterilization, packaging, storage, and any other handling process/stage that may detrimentally affect the stability and intended use of the device and thus jeopardize safe and efficacious function after implantation. Materials and devices that do not pass these prerequisite tests should not be evaluated for their biocompatibility.

9.3 METHODS FOR TESTING AND EVALUATING BIOCOMPATIBILITY

Evaluation of the biocompatibility of materials and devices consists of a sequence of tests and includes in vitro (using cells and tissues) tests, ex vivo (whenever applicable) tests, animal models, and clinical trials. Several guidelines and procedures have been developed for these purposes. Pertinent information can be found in the scientific and trade literature, and it can also be obtained from national and international standards organizations (such as the American Society for Testing and Materials, or ASTM), international organizations (such as the International Organization for Standardization or ISO), and federal agencies (such as the FDA and the National Institutes of Health, or NIH).

9.3.1 *In Vitro* Testing

In vitro tests have been used successfully to screen materials and devices for biocompatibility. Major advantages of this approach have been reasonable cost, small to reasonable capital investment in laboratory facilities and equipment, and, most important, relatively fast processing of large numbers of candidate materials and of device prototypes.

Blood compatibility (hemocompatibility) of materials and devices is determined by using anticoagulated blood (an unavoidable limitation of these tests) and evaluating formation of blood clots (thrombus formation) on their surfaces as well as activation of the plasma coagulation cascade, platelet adhesion, aggregation and concomitant synthesis and release of bioactive chemical compounds [such as aggregating agents and growth factors (described in Chapter 4)], activation of complement and of white blood cell (described in Chapter 6) when these constituents of blood interact with synthetic materials. Depending on the end use of the device, hemocompatibility tests must be conducted under either static or flow conditions during both acute and chronic tests. Damage of red blood cells leading to release of hemoglobin (hemolysis) under the flow conditions inside prosthetic devices as well as calcification associated with mechanically moving parts (such as leaflets of heart valves) must also be determined.

Unfortunately, these responses cannot be eliminated completely when blood comes in contact with synthetic materials; controlling and/or minimizing such responses has been the goal in the design of hemocompatible materials.

Advances in cell culture techniques have provided an exceptionally versatile and useful in vitro model for evaluating aspects of the biocompatibility of materials and devices pertinent to wound healing (discussed in Chapter 7). Mammalian cells of the tissue/organ relevant to a specific application can be used to determine cell functions (such as adhesion, migration, proliferation, synthesis and deposition of extracellular matrix chemical compounds, etc.) on the materials tested. Material properties that either support or prevent protein, and subsequent cell, interactions may be desirable. If either strong bonding between surrounding tissues and implanted materials or new tissue formation is the goal of the design and evaluation of the new materials tested, then only materials that support functions of specific cells and, concurrently, minimize interactions of competing cell lines should be evaluated further (as described in sections 9.3.2–9.3.3 of this chapter). For example, only materials that promote osteoblast function (the bone-forming cell) but minimize function of fibroblasts (competing cells) should become candidates for orthopedic/dental applications.

In vitro mammalian cell models can also be used to determine the effects of released chemical compounds (such as type and concentration of ions released during corrosion of metals as well as macromolecules and monomers released during degradation of bioresorbable and leaching polymers, respectively) under conditions that simulate those of the physiological milieu. Materials that fail the acute toxicity test should neither be evaluated nor considered further. Even when the material passes the cell viability test, the effects of released products on cell morphology (including intracellular accumulation of degradation products), proliferation, and other functions pertinent to the end use of the material and device must be evaluated.

Cellular in vitro models are now established tests for effective screening materials for biocompatibility. As with any other model, care should be taken in interpreting the results of these studies and in avoiding risky extrapolations. Undoubtedly, these models are very good for studying functions (and pertinent mechanisms) of one cell line at a time; such an approach, however, provides a limited perspective of the complex milieu of the body. Input from other models (for example, animal models) is needed to elucidate the multifaceted, interactive, and dynamic events that direct, mediate, and control tissue-biomaterial interactions inside the body.

9.3.2 Animal Models

9.3.2.1 Humane Concerns and Legal Regulations/Requirements

Animal models are used to determine the in vivo compatibility of materials and devices. Negative results seal the unacceptability of the tested systems. Unfortunately, positive results do not necessarily prove compatibility in humans. Because of species differences, extrapolation of conclusions from animal tests to predictions of human responses has proven unreliable and hazardous. Because of their homology to humans, the most desirable animal models are nonhuman primates; even then, species desirability must be balanced against issues such as the scarcity of these animals, the cost of purchasing and maintaining such colonies, etc.

Use of laboratory animals in research and testing bears unique responsibilities and should be considered only after successful completion of prerequisite material characterization tests (discussed in Section 9.2), appropriate computer simulation models, and of pertinent *in vitro* experiments (described in Section 9.3.1). Paramount among all concerns regarding the use of laboratory animals as experimental subjects is the recognized need to provide humane care and to prevent their exposure to unnecessary pain and suffering. Investigators, therefore, should identify the most appropriate species for the proposed study, carefully plan experiments using the smallest number of animals that will yield statistically useful information, and avoid unnecessary duplication of studies. Both investigators and sponsoring institutions must comply with local, state, and federal laws and regulations regarding animal experimentation. In the United States, the *Animal Welfare Act* (1985 and subsequent amendments) addressed care and use of laboratory animals, established laws, and set a system for monitoring, and thus controlling, animal research in the country. Compliance with these regulations by researchers who use laboratory animals is required by federal funding agencies such as the National Institutes of Health (NIH). Furthermore, animal research protocols must be reviewed and approved by an Institutional Animal Care and Use Committee (IACUC) before the study is initiated. It should be noted that these regulations identify protected animal species (exceptions, for example, include rats and mice bred for research purposes as well as livestock species that are used in agricultural research), as well as their procurement (from licensed suppliers), and specify requirements for husbandry, veterinary care, animal facilities, personnel training and credentials, experimental procedures (including physical restraint, anesthesia, and euthanasia), record keeping, and submission of appropriate reports at regular and specified time intervals. Organizations such as the American Association for Accreditation of Laboratory Animal Care and the American Association for Laboratory Animal Science provide pertinent information and accredit animal facilities at various institutions, whereas the American Society for the Testing of Materials makes available detailed protocols for testing materials and devices using animals. A series of publications by NIH (see the bibliography at the end of this chapter) has also provided helpful information on detailed protocols and pertinent methodology.

In summary, animal tests are a serious and major undertaking; these experiments are expensive, complex, and difficult to interpret, but they are also a useful and unavoidable precedent to human trials.

9.3.2.2 Classification of Tests

The animal tests used to evaluate material biocompatibility can be classified into the following three major categories:

Nonfunctional Tests
In this case, samples of arbitrary shape are implanted in soft tissues (that is, subcutaneously, intramuscularly, and intraperitoneally) via procedures that require minor surgery. Such studies are of relatively short duration (days to months) but provide valuable information regarding both local tissue-biomaterial interactions and systemic complications; these data are collected in the absence of mechanical loading and of any other pertinent functional operation of the prosthesis.

Ex Vivo Tests
Arterial-venous and venous-venous shunts are used to circulate blood from an animal, through the material(s) tested when inserted in the shunt loop before returning this blood back to the animal. In this case, protein accumulation, blood cell adhesion, and blood clotting on the material surface(s) are data used to determine the blood compatibility of the materials tested.

Functional Tests
These studies require implantation of a functional appropriately scaled version of the device, for example, a hip implant and a heart value in anatomic sites of animals under operational modes similar to those proposed for human use. Functional tests are long-term studies that require special considerations (such as design, fabrication, and testing of animal versions of prostheses) and are both complex and costly.

9.3.2.3 Local and Systemic Biological Responses

Implantation of materials and devices in animals provides valuable information regarding interactions with blood compatibility (described in Chapter 4), acute and chronic as well as local and systemic inflammation (described in Chapter 6), sensitization, and the overall associated wound healing process (described in Chapter 7). Pyrogenic, immunologic, toxic, and carcinogenic responses of animals to implant materials and devices can also be determined. Detailed methodologies and procedures for these tests are available in pertinent publications by ASTM, NIH, and various professional societies; examples of such references are provided in the bibliography at the end of this chapter. Table 9.3 illustrates a testing matrix based on the FDA and ISO guidelines; clearly the FDA requires a greater number of tests. To determine which tests should be conducted, the intended use (external, externally communicating, or internal), type of tissue contacted, and duration (transient, short-term, or long-term) of contact must be

TABLE 9.3. FDA/ISO test matrix

Device Categories			Initial Evaluation								Supplemental Evaluation	
	Body Contact	Contact Duration	Cytotoxicity	Sensitization	Irritation or Intracutaneous Reactivity	Systemic Toxicity (Acute)	Subchronic Toxicity (Sub-acute Toxicity)	Genotoxicity	Implantation	Hemocompatibility	Chronic Toxicity	Carcinogenicity
Surface Devices	Skin	A	●	●	●							
		B	●	●	●							
		C	●	●	●							
	Mucosal Membrane	A	●	●	●							
		B	●	●	●	○	○		○			
		C	●	●	●	○	●	●	○		○	
	Breached or Compromised Surfaces	A	●	●	●	○						
		B	●	●	●	○	○		○			
		C	●	●	●	○	●	●	○		○	
External Communicating Devices	Blood Path Indirect	A	●	●	●	●				●		
		B	●	●	●	●	○			●		
		C	●	●	○	●	●	●	○	●	●	●
	Tissue, Bone Dentin Communicating	A	●	●	●	○						
		B	●	●	○	○	○	●	●			
		C	●	●	○	○	○	●	●		○	●
	Circulating Blood	A	●	●	●	●		○		●		
		B	●	●	●	●	○	●	○	●		
		C	●	●	●	●	●	●	○	●	●	●
Implant Devices	Bone/Tissue	A	●	●	●	○						
		B	●	●	○	○	○	●	●			
		C	●	●	○	○	○	●	●		●	●
	Blood	A	●	●	●	●			●	●		
		B	●	●	●	●	○	●	●	●		
		C	●	●	●	●	●	●	●	●	●	●

A = Limited exposure (≤24 hours); B = Prolonged exposure (24 hours–30 days); C = Permanent contact (>30 days)
● = FDA and ISO evaluation tests; ○ = Additional tests required by the FDA

Adapted from: Anand, V.P., "Biocompatibility safety assessment of medical devices: FDA/ISO and Japanese Guidelines", *Med. Dev. Diagn. Ind.* 22:206–219, 2000.

specified. As would be expected, more extensive testing is required for permanently implanted devices.

9.3.3 Clinical Trials

9.3.3.1 Procedures and Regulations

No matter how successful the in vitro and animal tests are, it is not possible to predict the performance of devices in humans without clinical trials. In fact, successful clinical trials are required before a biomedical device becomes available to the general public, that is, patients who would be recipients of these prostheses.

In addition to scientific standards and criteria, clinical evaluations of biomedical devices must comply with legal regulations. Carefully documented and successfully completed pertinent in vivo, ex vivo, and in vivo (animal) tests are needed to support applications to federal regulatory agencies such as the Food and Drug Administration (FDA) for clinical trials. A strong case of the benefits (as well as assurance of the absence of unusual risks) to recipients of the new device must be made. In addition, detailed protocols describing the device, surgical procedure, postoperative treatment and medical care of recipients, and evaluation [for example, comparisons of patient condition/welfare before and after implant surgery, comparison of patients/recipients with healthy humans in appropriate control (age, sex, living and/or working environment) groups, etc.] must be reviewed by Institutional Review Boards and comply with guidelines set by the FDA and by international regulations (such as the *Declaration of Helsinki*) that protect and ensure the rights and dignity of human subjects participating in clinical trials.

With FDA approval, biomedical devices (consisting of biomaterials) can then be implanted into a small number of human subjects as part of limited, but carefully controlled and closely monitored, studies that involve a sequence of three phases (designated by the numerals I, II and III). Patients must give their informed consent before they participate in these trials. Ethical considerations come into play in making crucial decisions such as who will be the patients either rejected or chosen to receive a device that may save, prolong, and/or improve the quality of their lives. The identity and medical records of participating patients must remain confidential forever. Reports on the successful outcome of the clinical trials (including positive and adverse results, patient follow-up, quantification of data, and discussion of the significance of the conclusions) must be submitted to the FDA for final approval of a device prior to marketing. Undoubtedly, inventors and companies must conduct multifaceted and demanding scientific studies as well as investing lots of time and money to reach this point in the development of marketable biomedical products.

9.3.3.2 *Implant Failure, Retrieval, and Evaluation*

Different, but important and valuable, information can be obtained from evaluating materials and devices, as well as surrounding biological tissues, on retrieval of prostheses from the bodies of humans and animals at the end (unrelated to implant failure) of the recipients' lives.

It should be kept in mind that, although biomedical material and device design is based on physiological data from normal, healthy individuals, the recipients of these systems are primarily either elderly or ill people. It is impossible to predict device performance under all possible scenarios of the milieu of the body of patients in the presence of various diseases and prescribed medications. Under these circumstances unanticipated complications, such as functional or physicochemical material failure of implants, may occur. In addition to pain and discomfort, such complications may be associated with infection and other clinical symptoms that jeopardize the welfare of patients. Under these circumstances, surgical intervention and retrieval of implants is unavoidable.

Examination of implants retrieved for reasons unrelated to function of the prostheses provides unquestionably useful evidence regarding the safety and efficacy of the device. Detailed study of failed implants is needed to determine the cause(s) of failure. This information could, in turn, be used to improve design and fabrication of materials and devices, establish selection criteria for implants, and develop revised protocols and techniques for various stages of the evaluation of implants. In addition to improving current knowledge of the events taking place at the tissue-implant interface, such developments could lead to the clinical success of biomedical prostheses.

9.4 ATTEMPTING TO DEFINE BIOCOMPATIBILITY

The term *biocompatibility* has been defined in many and different ways. Historically, materials that caused minimal biological responses were considered "biocompatible." Thus, a material was considered acceptable for implantation if it did not provoke adverse host reactions, resulted in minimal inflammation, was nonimmunogenic, etc. The "old" definition of biocompatibility placed exclusive emphasis on the response of the body to the biomaterial. With the exception of cases of drastic material degradation and consequent release of harmful byproducts that subsequently elicited adverse biological responses *in vivo*, material responses were not explicitly addressed in the original concept of biocompatibility.

Biocompatibility, therefore, must account for the interactions between tissues and biomaterials. The use of the prefix "inter" implies that not only can biomaterials affect biological responses, but also that the milieu of the body can affect materials. Thus, whether a material is biocompatible depends on suitable host *and* on material responses in a specific application or, in other words, whether the material performs as intended.

9.5 CONCLUDING REMARKS

Scientific opinion regarding sufficient and necessary criteria necessary for successful implant material and devices has been changing, reflecting progress made in "old" fields of research and inroads in new directions. In this respect, then, it is not surprising that past attempts to define biocompatibility in either broad terms (for example, success in a specific application) or in terms of what a material and device should not be (that is, nonpyrogenic, nontoxic, nonallergenic, nonthrombogenic, etc.) are either limited or dated. Currently, cellular, molecular, and genetic level research is producing exciting new information that has improved understanding of the host/biomaterials interactions. Application of this knowledge could lead to development of new generations of biomaterials and implant devices as well as to improved strategies for their evaluation and use. Proactive biomaterials, that is, materials that elicit specific, desired, and timely responses from surrounding cells and tissues in the bodies of humans and animals are the future of this field of scientific endeavor and engineering applications.

9.6. SUMMARY

- Biomedical implants (as well as their constituent materials) must be evaluated in a series of tests of increasing complexity in order to establish safety and efficacy (biocompatibility).
- In vitro tests are used to screen materials and devices before proceeding to animal tests.
- In vivo tests must address the intended application of each device.
- *In vitro*, *ex vivo* and/or animal experiments provide valuable information but do not definitively predict the performance of devices in humans.
- Successful clinical trials are required before a biomedical device becomes commercially available to patients.
- Various agencies (such as NIH) and professional organizations (such as ASTM and ISO) have compiled detailed instructions regarding protocols and methods pertinent to each stage of the biocompatibility evaluation process of materials and devices.
- All biocompatibility evaluations should be conducted in accordance with institutional regulations and must comply with pertinent federal and international laws.

9.7 BIBLIOGRAPHY/SUGGESTED READING

Hanson, S., Lalor, P.A., Niemi, S.M., Northrup, S.J., Ratner, B.D., Spector, M., Vale, B.H., and Willson, J.E. Chapter 5: Testing Biomaterials. In *Biomaterials Science: An*

Introduction to Materials in Medicine, Ratner, B.D., Hoffman, A.S., Schoen, F.J., Lemons, J.E., (eds.), Academic Press, New York, NY (1996), pp. 215–242.

Harker, L.A., Ratner, B.D., Didisheim, P. (eds.), Cardiovascular Biomaterials and Biocompatibility: A Guide to the Study of Blood-Tissue-Material Interactions, Supplement to *Cardiovascular Pathology*: 2 (1993).

National Institutes of Health. *Guidelines for Physicochemical Characterization of Biomaterials.* Publication 80-2186. U.S. Department of Health and Human Services, Washington, D.C. (1980).

National Institutes of Health. *Guide for the Care and Use of Laboratory Animals.* Publication 86-23. U.S. Department of Health and Human Services, Washington, D.C. (1985).

National Institutes of Health. *Guidelines for Blood-Material Interactions.* Publication 85-2185. National Institutes of Health, Bethesda, MD (1985).

National Institutes of Health. *Public Health Service Policy on Human Care and Use of Laboratory Animals.* Office of Protection from Research Risks. National Institutes of Health, Bethesda, MD (1986).

9.8 QUIZ QUESTIONS

1. You are the biomedical engineer responsible for evaluating the compatibility of products of your company.
   Discuss in detail the process you will follow while evaluating the blood compatibility (hemocompatibility) of a new product, a vascular prosthesis.

2. A company is considering a new metal for orthopedic implants. You are the biomedical engineer responsible for evaluating product biocompatibility.
   (a) Outline the process you will follow in completing this assignment.
   (b) Give pertinent details of the process you described in part (a).
   Discuss advantages *and* limitations of the approaches you plan to use.
   (c) Identify unique aspects of the system under consideration. Discuss special concerns *and* how you will resolve them.

3. The company where you work is considering a new, nonporous polymer for vascular grafts. You are the biomedical engineer responsible for evaluating biocompatibility of materials and devices.
   (a) Outline the process you will follow in completing this assignment.
   (b) Give pertinent details of the process you described in part (a). Discuss advantages and limitations of the approaches you plan to use.
   (c) Identify unique aspects of the system under consideration. Discuss special concerns *and* how you will resolve them.

4. Discuss the advantages, disadvantages, and pertinent special considerations inherent in the design and use of (a) "functional" and (b) "nonfunctional" in vivo animal tests.

9.9 STUDY QUESTIONS

1. You are a biomedical engineer working in the Research and Development Division of a company in the cardiovascular prostheses field. Your project is to design the *ideal* blood compatible material for vascular prostheses.
   (a) What special properties/characteristics should this material have? Justify your choices and criteria with appropriate explanations and with pertinent arguments.
   (b) Discuss the features you will incorporate in your design in order to *enhance* the blood compatibility (hemocompatibility) of the novel biomaterials.

2. You are the biomedical engineer responsible for evaluating the biocompatibility of hip prostheses, a product of your company.

   Discuss (in appropriate detail) the tests you will conduct to establish the biocompatibility of these products and thus ensure that they are ready for the clinical trials.

3. Discuss the advantages, disadvantages, limitations, and ethical issues pertinent to
   (a) in vivo animal tests, and
   (b) clinical trials
   used to evaluate the biocompatibility of implant materials and devices.

10

Example 1. Opening Occluded Vessels: Vascular Grafts, Intimal Hyperplasia*

Arteriosclerosis is a chronic disease marked by thickening, hardening, and stiffening of the arterial walls, which results in impaired blood circulation. Atherosclerosis is one form of arteriosclerosis, in which plaques containing cholesterol and lipids are deposited inside arteries, blocking blood flow. Severe impairment of blood flow requires surgical intervention, for example, removing the blocked portion of the artery and replacing or bypassing it with a *vascular graft*—a new, unoccluded segment of vessel; inserting a stiff, annular framework (a *stent*) inside the artery to hold the vessel open; or performing a balloon angioplasty procedure to flatten the obstruction and clear the center of the vessel.

Autologous vessels (autografts) are the preferred replacement for diseased or damaged blood vessels, but synthetic vascular grafts are often needed because of unavailability or inadequacy of autografts. Expanded polytetrafluoroethylene (ePTFE) is the most widely implanted material for synthetic vascular grafts. Knitted poly(ethylene terephthalate) (Dacron) grafts are also used. After implantation, fibrous tissue encapsulates the outside of these grafts and can grow into their microporous surfaces. The inner surface of the grafts becomes covered with a "pseudointima" consisting of fibrin and fibroblasts. This pseudointima is essentially the blood-biomaterial interface, because endothelial cells rarely populate the surfaces of these grafts.

Although grafts made from synthetic biomaterials perform well when used to replace larger blood vessels, they are inadequate for replacing small diameter (<4 mm) vessels. Consider the effect of formation of a pseudointima, or a thrombotic mass, on the cross-sectional area of the *lumen* (the inner, open space of a tube) of a small-diameter graft. Deposition of 1 mm of biological material on the inner surface of a graft with a 10-mm inner diameter results in a 36% decrease in the cross-sectional area available for blood flow. In contrast, 1 mm of pseudointima/thrombus deposited in a 4-mm-inner diameter graft decreases

*A first draft of this chapter was written by Amanda Filanowski, Tulane University.

the available cross-sectional area by 75%. The significantly reduced cross-sectional area gives a greater resistance to blood flow and thereby slows the flow through the vessel. This, in turn, allows platelets and circulating elements of the coagulation cascade (clotting factors, etc.) to interact with the pseudointima for a longer period of time than would occur in a large-diameter graft. The increased residence time and blood-material interactions can cause formation of clots that, even though relatively small in size, can completely occlude the smaller grafts. Significant resources are being devoted to development of materials for use in small-diameter synthetic vascular grafts, especially for the replacement of blood vessels below the knee as treatment for trauma or clinical conditions marked by chronically reduced circulation in the extremities.

If the occluded blood vessel is one of the major vessels supplying the heart (one of the coronary arteries), the cardiac surgeons and the patient may consider using a vascular graft to route blood flow around the occlusion (bypass surgery). A *balloon angioplasty* procedure, in which a catheter equipped with a tiny balloon at the tip is inserted into the narrowed portion of the artery, may also be an option. The balloon is inflated, pressing the accumulated plaque back against the vessel intima, and thus widening the lumen of the artery. The balloon is then deflated, and the catheter is withdrawn.

Both bypass surgery and balloon angioplasty can significantly improve the quality of life of a patient who has suffered from arterial blockage. However, after 10 years, 44% of bypass autologous saphenous vein grafts and 50% of coronary angioplasty procedures fail because of intimal thickening, luminal narrowing, and thrombosis [1]. A primary cause of this long-term vessel thickening and narrowing is a process called *intimal hyperplasia.*

Intimal hyperplasia is an exaggerated wound healing response to arterial injury, which results in an overaccumulation of cells and extracellular matrix within the intima—reblocking the lumen of the vessel. Acute causes of arterial injury can include the dissection and suturing that takes place during surgery or the pressures applied during an angioplasty procedure. Blood vessels can also undergo chronic injury over long periods of time, caused by, for example, changes in the hemodynamic forces associated with blood flow, nicotine exposure, hypertension, and diabetes. The acute injuries listed above can cause the endothelial cells of the intima to be injured, stripped away, or squashed underneath an angioplasty-flattened plaque. The cells in the *media* portion of the blood vessel, underlying the intima, are stretched as the blood vessel is manipulated or expanded during angioplasty. Blood contact with the damaged tissues and cells initiates platelet aggregation, coagulation, and inflammation, as discussed elsewhere in this textbook. Moreover, smooth muscle cells in the media and myofibroblasts in the adventitia begin to proliferate 1–2 days after an acute injury. This cell proliferation is a necessary part of the vessel wound healing process. However, in intimal hyperplasia, these cells rapidly overproliferate and migrate from the media into the intima. The cells can then continue to actively proliferate and produce an extracellular matrix that includes collagen, elastin, proteoglycans, and sometimes lipids, for weeks. A new, steady-state tissue arrangement may not be reached until 3 months after injury. By this time, a new (neo) intima has been formed from the resulting combination of cells and matrix. The loss of the normal endothelial cell lining causes that portion of the blood vessel to lose much of its normal anticoagulant properties. Clots and plaques are more likely to form on the neointima than on a normal, healthy endothelial intima. If the thrombi or plaques break away from the neointimal surface and travel through the vasculature to occlude a smaller blood vessel, the result can be acute myocardial infarction (heart attack) or stroke, and sudden death. If the thrombi or plaques remain on the neointimal surface, they may grow and thicken to eventually reocclude the vessel.

Intimal hyperplasia is a complex problem, and the underlying causative mechanisms are not yet fully understood. A number of factors (including biochemical and mechanical consequences of injury and wound healing, altered blood flow and hemodynamics, and chemical and mechanical interactions of an implanted biomaterial with surrounding tissue) are probably involved, to some degree, in the development of intimal hyperplasia. Relevant questions to ask at this point are: Is intimal hyperplasia more a function of acute injury circumstances or of chronic conditions? How is the proliferation of cells normally limited within blood vessels? Do vascular graft materials and normal arteries transmit the same mechanical stresses to surrounding cells and tissues? Are areas of irregular blood flow more likely to experience intimal hyperplasia?

Acute injuries, such as denuding (stripping away) the endothelial lining of the intima, remove the natural interface between the blood and the vessel wall. The normal, healthy endothelium is the source of a number of chemical agents that prevent blood coagulation, control vascular tone, prevent or promote epithelial and smooth muscle cell proliferation, induce inflammation, and degrade extracellular matrix. Removal of the endothelium may be enough of a stimulus to induce intimal hyperplasia. Endothelial cells are extremely sensitive to a wide variety of injury mechanisms, and after injury the functions of these cells can change dramatically. For example, damaged endothelial cells release basic fibroblast growth factor, which stimulates smooth muscle cell proliferation, and platelet-derived growth factor, which stimulates smooth muscle cell migration and can prevent *apoptosis*, or programmed cell death—perhaps accounting for the increased cell numbers of the intima. It has been hypothesized that the healthy endothelium acts as a barrier to growth factors and biochemicals that would otherwise promote proliferation of medial and adventitial cells, so a compromised endothelium permits cell overgrowth. Healthy endothelial cells also release various chemical agents that have been shown to inhibit smooth muscle cell proliferation in vitro (for example, nitric oxide and heparin sulfate). If injured endothelial cells lose the ability to release these chemical mediators, cell overgrowth could occur in a chronic fashion, long after the initial injury to the endothelium. Finally, repeated injury can cause endothelial cells to lose the ability to replicate. If the cells cannot reproduce, denuded areas of the subendothelium will be left exposed to the blood, permitting coagulation and inflammation and encouraging the accumulation of clots, plaques, etc.

Do vascular graft materials and normal arteries transmit the same mechanical stresses to surrounding cells and tissues? If not, how might that affect the development of intimal hyperplasia? The short answers to these questions are "probably not" and "it's not yet clear." Implanting a vascular graft involves the creation of *anastomoses*, or connections between two blood vessels. The vessels can be connected end-to-end, forming a continuous straight path for blood flow, or end-to-side, making the blood flow through some form of curved or angled vessel. Intimal hyperplasia is often quite pronounced around and near anastomotic sutures, perhaps because the vascular graft/arterial interface can alter both the local fluid mechanics (blood flow) and solid tissue mechanics (strains in the graft and surrounding tissues). Irregular blood flow patterns (such as flow stagnation, separation, and recirculation) that lead to long fluid residence times and altered fluid shear stresses exerted on the cells of the intima have both been linked to the development of intimal hyperplasia.

Compliance mismatch is also a concern in bypass surgeries. The mechanical properties of the vascular graft material—whether the graft is composed of a synthetic material, a portion of a vein, or even a portion of another artery—are often different from that of the native vessel. This can result in increased stresses and strains at the graft/vessel interface, as well as altered vessel geometry and blood flow patterns. It is conventional wisdom that graft *patency* (success) rates increase with better compliance matches between the graft and the artery. Depending on the constitutive material, synthetic grafts tend to be stiffer than arterial tissues. Arterial grafts provide the best possible compliance match for coronary bypass surgery, which is one reason why the internal thoracic artery is considered a better autograft than the saphenous vein. The saphenous vein is still the most commonly used vascular autograft, however, because a typical coronary bypass patient possesses few healthy, undiseased arteries. Compliance mismatch between venous grafts and arteries occurs because of the differing hemodynamic forces that each experience in situ. The arterial circulation is subject to greater tangential wall stress (i.e., blood pressure) than the venous system. Arteries are generally thicker than veins and contain more muscle and elastic tissue than veins. At any given time, approximately 15% of the total blood volume is in the arterial circulation at an average pressure of 100 mm Hg, while approximately 60% of the total blood volume is in the venous circulation at an average pressure of only 10 mm Hg [3]. It has been shown that intimal smooth muscle cell proliferation in vein grafts will stop when the ratio of graft radius to graft wall thickness matches the ratio found in arteries [4], essentially normalizing shear and wall stresses between the two types of vessels. This active remodeling of the venous graft structure is a response of the constituent cells to the altered blood flow characteristics in the arterial environment. Researchers are investigating the possibility of preconditioning venous grafts by exposing them to arterial flow conditions before implantation, to reduce the severity of compliance mismatch and thus reduce the chance of resulting neointimal growth and intimal hyperplasia.

SUMMARY

Opening occluded blood vessels can significantly improve the quality of a patient's life. Currently, even the very best methods and tools for removing an occlusion, or for providing a route for blood to flow around an occlusion, still have room for improvement. Biochemical and biomechanical aspects of blood-biomaterial and cell-biomaterial interactions are crucial to the long-term success of a reopened blood vessel, under the dynamic and demanding conditions that exist in vivo.

REFERENCES

[1] Allaire, E. and Clowes, A.W., "The Intimal Hyperplastic Response," *The Annals of Thoracic Surgery*, 64:S38–S46, 1997.

[2] Allaire and Clowes, p. S39.

[3] Vander, A., Sherman, J., and Luciano, D., *Human Physiology: The Mechanisms of Body Function*, (Boston: WCB McGraw-Hill, 1998) p. 423.

[4] Allaire and Clowes, p. S40.

SUGGESTED READING

Clowes, A.W., "Pathologic intimal hyperplasia as a response to vascular injury and reconstruction," in *Vascular Surgery*, fifth edition, R.B. Rutherford, ed., W.B. Saunders Company, Philadelphia, PA, 1999.

Greenwald, S.E. and Berry, C.L., "Improving vascular grafts: the importance of mechanical and haemodynamic properties," *Journal of Pathology*, 190:292–299, 2000.

Sauvage, L.R., "A brief history of arterial prosthesis development," *Journal of Investigative Surgery*, 6:221–225, 1993.

Szilagyi, D.E., "Arterial substitutes: thirty years of success and failure," *Annals of Vascular Surgery*, 1:357–363, 1986.

DISCOVERY ACTIVITIES: STENT CASE STUDIES

1a. Make a list of issues that must be considered in the design of stents, and propose general solutions for each issue. You may want to start your list of issues by considering the interface between the stent and the flowing blood, as well as the interface between the stent and the arterial wall. How can a stent be placed into position with a minimally invasive procedure? How should this device be sterilized and packaged before use?

1b. Now find at least two different stents currently on the market, and see how your list of design issues compares with these products. What biomaterials are used to make stents? How do these materials interact with blood and with the arterial wall?

1c. For undergraduate students considering biomedical careers: What companies design and manufacture stents? From the initial design and testing to the final sales and use of the stents, how many different types of employees/job positions are involved in providing a safe and effective product for use in patients? Which of these positions would you consider pursuing someday? What qualifications would you need to obtain?

2. Part of the text above states: "After implantation, fibrous tissue encapsulates the outside of these grafts and can grow into their microporous surfaces. The inner surface of the grafts becomes covered with a "pseudointima" consisting of fibrin and fibroblasts. This pseudointima is essentially the blood-biomaterial interface, because endothelial cells rarely populate the surfaces of these grafts." On the basis of the information you've learned from this book, and thinking on the cellular and molecular levels, explain why and how the pseudointima forms. What are some benefits and drawbacks of this blood-biomaterial interface compared with a layer of living endothelial cells?

3. It was stated in the above text that the prevention of apoptosis, or programmed cell death, might result in an overpopulation of cells. What is this "programmed cell death," and how is it different from cell death due to injury, disease, or age?

4. Research surgical suturing techniques used in implanting vascular grafts (suture biomaterials, suture patterns, stapling, etc.). Are any of these fastening methods and biomaterials advocated for reduction of intimal hyperplasia?

5. It would seem that, if a healthy endothelium is crucial for the long-term patency of blood vessels, vascular grafts should be seeded with living endothelial cells before implantation. Perform a literature search to learn about efforts to develop such grafts. Why isn't this biotechnology already commercially-available?

6. It is known that endothelial cells actively respond to fluid shear stresses (for example, from normal or altered blood flow). Look into the scientific literature. How are endothelial cell responses to fluid shear experimentally investigated in the laboratory? What are some major findings of these experiments? How are endothelial cell responses to fluid shear investigated with computational modeling techniques? What are some major predictions of these models?

10

Example 2. Replacing Joints and Teeth

Musculoskeletal impairments and injuries affect an estimated 60 million people in the United States alone. As a result of these conditions, over 400,000 hips and knees are replaced each year in the U.S. Furthermore, between 100,000 and 300,000 dental implants are used each year to replace teeth lost because of caries (tooth decay) or trauma. Whereas the success rate of dental implants is up to 95% at 15 years after implantation, more than 60,000 hip and knee implants are revised each year.

After implantation of metallic joint and dental implants, a complex series of events takes place, ideally leading to *osseointegration*. This term was coined by Dr. Per-Ingvar Brånemark and colleagues, who conducted landmark research on dental implants. As the two parts of the word imply, osseointegration refers to incorporation of an implant in bone. More specifically, the implant is integrated into the bone without an intervening layer of fibrous tissue. This integration has functional significance, in that the device can bear the loads imposed on it during normal physiological use—whether that use is chewing on a dental implant or walking on a hip implant.

For osseointegration to occur, however, certain requirements must be met. For example, gaps between bone and implant must be bridged, and any bone that was damaged during preparation of the implant site must be repaired. During this time, unfavorable conditions that can disrupt the newly forming tissue must be prevented. In addition to implant design-related factors (such as material, shape, topography, and surface chemistry), the degree of osseointegration is affected by numerous factors that biomaterials engineering cannot control, including mechanical loading and surgical technique, as well as patient variables such as bone quantity and quality.

Excessive interfacial *micromotion* during bone healing is detrimental to osseointegration. Although an oversimplification of its effects, relative motion between bone and an implant (micromotion) damages the fibrin network and new vasculature that are part of the early bone healing process, consequently rerouting the healing response into repair by scar tissue. To reduce/prevent micromotion, lengthy periods of restricted loading are prescribed postoperatively.

For example, total hip arthroplasty patients can be protected from full weight bearing for at least 6–12 weeks, and in some cases for up to 6 months. To promote osseointegration of dental implants, a two-stage procedure is recommended in which a 3- to 6-month healing period separates implantation of the *fixture* (the part embedded in bone) from attachment of the *abutment* (a postlike part to which the artificial tooth is attached). Delayed implant loading is required because it normally takes 6 weeks for woven callus to mature and 18 weeks for lamellar organization of the newly formed bone and development of adequate load-carrying capacity [1].

Meticulous surgical procedures are required to obtain osseointegration. An increase in bone temperature to just 47 °C (for example, during drilling or polymerization of bone cement) can at least partially inhibit bone formation [2, 3]. Other important factors that influence stability of the bone-dental implant interface include the fit of the device in the implantation site and the amount and health of bone surrounding an implantation site. Adequate bone quantity and quality, which may not be present in all patients, are needed to accommodate and support the implant. For example, integration of implants in patients with osteoporosis or diabetes is different than in patients without conditions that affect the musculoskeletal system. Also, bone has a limited ability to spontaneously span gaps between bone and implant. Gaps that are more than 1–2 mm can compromise the strength of attachment and bone ingrowth around implants. Because of geometric differences between the manufactured implant and implantation site, precise fit is often difficult to achieve.

Even in the case of osseointegrated implants, when there is no fibrous capsule, high-resolution microscopic studies have revealed an afibrillar interfacial zone at the bone-implant interface; mineralized tissue generally does not directly touch the biomaterial. Although its thickness and appearance vary, this zone forms regardless of the type of biomaterial implanted, including commercially pure (cp)Ti, stainless steel, and hydroxyapatite. The interfacial layer is rich in noncollagenous proteins as well as certain plasma proteins. Some researchers have suggested that this interfacial zone provides a mechanism for "bonding" between natural hard tissue and biomaterial, but its mechanical weakness makes this explanation unlikely.

Exposure to a simulated or actual biological environment causes significant changes in the surface of a dental/orthopaedic implant. Continued oxidation of metallic implants occurs both in vivo and in vitro. Although cpTi, Ti-6Al-4V, and Co-Cr-Mo are often selected as dental/orthopaedic biomaterials because of their stable oxide films, they still undergo electrochemical changes in the physiological environment. For example, cpTi implants have an oxide thickness of 2–6 nm before implantation and films on implants retrieved from human tissues are two to three times thicker. Furthermore, surface analyses show that Ca, P, and S from the physiological environment are incorporated and change the chemical composition of the oxide film.

In an effort to control tissue-implant interactions and promote osseointegration, several biomaterial surface modification strategies have been investigated.

An objective is to design implants that will promote osseointegration, regardless of type of implant, implantation site, early loading, bone quality, etc. The surface modification approaches can be classified as physicochemical, morphologic, or biochemical.

Surface energy, surface charge, and surface composition are among the physicochemical characteristics that have been altered with the aim of improving the bone-implant interface. For example, glow discharge is a process in which energetic species in an ionized gas "scrub" adsorbed contaminants and other molecules from the surface, and thereby increase surface free energy, with the aim of increasing tissue adhesion. One of the most widely studied physicochemical modifications is deposition of calcium phosphate coatings. Because of the chemical similarity to bone mineral, it is hypothesized that bone can bond to these coatings. Calcium phosphate-coated hip and dental implants are commercially available.

Alterations in surface morphology and roughness have been used to influence cell and tissue responses to implants. Porous coatings were developed on the basis of reports from the 1970s that extensive bone ingrowth occurred when pores were at least 100 μm in diameter. For smaller pores, fibrous tissue occupies the void space because formation of an extensive capillary network, needed for *osteogenesis*, is impaired. The mechanical interlocking that results from ingrowth of bone into porous coatings can stabilize implants. In addition to providing mechanical interlocking, surfaces with grooves can induce *contact guidance*, whereby the direction of cell movement is affected by the morphology of the substrate. This phenomenon has applications in preventing epithelial downgrowth on dental implants and perhaps in directing bone formation along particular regions of an implant.

A newer category of surface modifications involves changing the biochemical properties of surfaces. In biochemical surface modification, peptides or proteins are immobilized on biomaterials to control cell and tissue responses. In contrast to changing the physicochemical properties of a surface, such as by depositing a calcium phosphate coating, this approach utilizes critical organic components of bone to affect tissue behavior. For example, attachment of cell adhesion signals to surfaces has received extensive interest. Various peptides containing the ubiquitous Arg-Gly-Asp (RGD) sequence have been immobilized for cell adhesion mediated by cell surface receptors in the integrin superfamily. Because of redundancy in the affinity of integrins for adhesive proteins and because a variety of cells possess the same integrins, the challenge is to design peptide-modified surfaces that are specific for selected cell types, like osteoblasts. Researchers are attempting to circumvent this problem by using longer peptides having different conformations or even non-RGD peptides. A second approach to biochemical surface modification involves attachment of biomolecules that affect osteoblast growth, activity, and differentiation, such as those responsible for bone development and fracture healing. Many bone growth factors are readily available for this use because of recombinant DNA technology. The challenge for effectively using this approach, however, is to place enough of the protein(s) at the tissue-implant interface for a period sufficient to promote the desired cell responses and to keep the protein(s) in a biologically active state long enough to initiate the responses. Methods being investigated to control exposure and concentration of bone growth factors at the cell-biomaterial interface include adsorption, covalent immobilization, and controlled release from coatings. At present, implants modified by these methods are not often commercially available. Biochemical surface modification will likely find application in combination with other surface modification approaches.

SUMMARY

Replacement of diseased and damage joints and teeth can significantly improve the quality of a patient's life. Dental implants, placed in a two-stage procedure with meticulous surgical technique, can last for more than 15 or 20 years. The success of hip and knee implants, however, is not always as impressive. Alterations in the surface properties of implants and control of micromotion at the tissue-implant interface are likely to improve implant performance.

REFERENCES

[1] Roberts, W.E., "Bone tissue interface," *J. Dent. Educ.* 52:804–809, 1988.

[2] Eriksson, R.A. and Albrektsson, T., "The effect of heat on bone regeneration," *J. Oral Maxillofac. Surg.* 42:701–711, 1984.

[3] Wykman, A.G., "Acetabular cement temperature in arthroplasty. Effect of water cooling in 19 cases," *Acta Orthop. Scand.* 63:543–544, 1992.

BIBLIOGRAPHY/SUGGESTED READING

Cowin, S.C., ed., *Bone Mechanics Handbook*, CRC Press LLC, Boca Raton, FL (2001).

Martin, R.B., Burr, D.B., and Sharkey, N.A., *Skeletal Tissue Mechanics*, Springer-Verlag, New York, Inc. (1998).

Puleo, D.A. and Nanci, A., "Understanding and controlling the bone-implant interface," *Biomaterials* 20 (1999): 2311–2321.

DISCOVERY ACTIVITIES

1. Draw a vertical line down the middle of a piece of paper. If this line is the interface between a bone and a metallic dental/orthopedic implant, create an instructional

schematic (that students could use as a study guide) depicting the implant material responses to the biological environment (on one side of the line) and the bone cell and tissue responses to the presence of the implant (on the other side of the line).

2. Explore the work of Dr. Per-Ingvar Brånemark—Why is his work important? What types of dental implants use a two-stage implantation procedure today? What types of surface treatments are performed on commercially available dental implants to improve osseointegration? What types of surface treatments are currently being researched and developed to improve osseointegration of dental implants?

Orthopedic Implant Case Studies

3a. How can a hip or knee be securely placed into position while causing minimal trauma and damage to the surrounding tissues? If you can arrange with a surgeon to watch a hip or knee replacement take place, do so; otherwise, find a videotape or a television show that focuses on operations and watch a knee/hip replacement. What types of trauma are involved in these implantation processes? What portions of the wound healing process will be strongly activated, and how might these affect the bone-implant interface?

3b. Find at least two different hip implants currently on the market. What biomaterials are used to make these implants? Are the surfaces of the implants specially designed or modified for improved bone biocompatibility—and if so, how?

3c. For undergraduate students considering biomedical careers: What companies design and manufacture knee or hip implants? From the initial design and testing to the final sales and use of the implants, how many different types of employees/job positions are involved in providing a safe and effective product for use in patients? Which of these positions would you consider pursuing someday? What qualifications would you need to obtain?

4. There is a great deal of scientific literature on the responses of bone cells and tissue to mechanical stimuli. Perform a literature search to address one of these topics: What is Wolff's law, and how might this "law" be applied to tissue modeling and remodeling at the bone-implant interface? List and describe the different techniques used in laboratories to expose cells to different types of mechanical stimuli. How are cells (in general) thought to experience different types of mechanical stimuli—What types of receptors are involved, and how can a cell "respond" to a mechanical stimulus? How might cells in bone tissue experience fluid shear stresses, and how does fluid shear affect the functions of bone cells? If bone cells are cultured on a surface, and that surface is deformed or strained, how does the strain affect the functions of bone cells?

Design Challenge: the Ultimate Bone-Biomaterial Interface

5a. You've been challenged to design the "ultimate" biomaterial for use in dental/orthopedic applications. What should some of the bulk properties of this ultimate material be? What cellular and molecular level events at the bone-implant interface would the surface of this ultimate material be designed to encourage? What events would the surface be designed to discourage?

5b. Which of the desirable interfacial (surface) events that you listed for question 5a would you prioritize as being the most important for your ultimate biomaterial to control? How do you justify your decision?

5c. Keeping in mind the interfacial event you chose as the most important for question 5b, obtain and read the paper by Puleo and Nanci listed in "Suggested Reading." How could you use physicochemical methods to create a biomaterial surface that would encourage that event? How could you use morphologic methods? How could you use biochemical methods? Given a choice, would you choose one of these three general approaches or a combination of the three to design your ultimate biomaterial?

5d. In what ways could the functionality of your ultimate biomaterial surface be altered or compromised during implantation and the initial stages of the wound healing process?

5e. What types of in vitro experiments would you want to do to test the functionality of your ultimate biomaterial? What types of in vivo experiments would you do? What would the experimental hypotheses be? What would you measure (and how could you perform the measurements) to generate the data that would be used in testing your hypotheses?

Answers to Quiz Questions

CHAPTER 1

1. Any natural or synthetic material, other than a drug, used to repair, augment, or replace a tissue or organ of the body.

2. A material consisting of electropositive elements in which the positively charged ion cores are immersed in an electron "cloud." Commercially, pure titanium has been used quite successfully for dental implants.

3. A material that is a compound of metallic and nonmetallic elements. Hydroxyapatite has been used as a coating on hip implants to facilitate bonding between bone and the prosthesis.

4. A material consisting of macromolecules composes of repeating units. Ultrahigh-molecular-weight polyethylene has been used as an articulating surface in joint replacements.

5. Thermal and mechanical processing of materials change the arrangements of constituent atoms.

6. Degradation can compromise the mechanical properties of the implant as well as release by-products (such as ions, chemical compounds, and particulate debris) that may cause adverse biological responses.

CHAPTER 2

1.
```
      COOH
       |
  H2N-C-H
       |
       R
```

2. Sequential condensation reactions that result in peptide bonds between amino acids.

3. The α-helix is a right-handed helix stabilized by hydrogen bonds between approximately every fourth amino acid. The β-strand is an extended zigzag chain, which can be hydrogen-bonded to neighboring β-strands to form a β-sheet.

4. Hydrogen bonds—disrupted by increased temperature
   Ionic bonds—disrupted by increased salt concentration or changes in pH
   Hydrophobic interactions—disrupted by chemicals (e.g., chaotropes) that alter water structure
   Covalent bonds—peptide bonds cleaved by enzymatic action or disulfide bridges disrupted by reducing chemicals

5. The quarter-stagger arrangement of tropocollagen molecules. Because the gaps in the array do not bind heavy metals, which are used to provide contrast in the tissue, these regions are less electron dense and allow electrons to pass through more easily, resulting in bright regions in the image. Conversely, regions of overlapping tropocollagen molecules are more electron dense and appear as dark bands.

6. A large number of intra- and intermolecular cross-links.

7. Arginine-glycine-aspartic acid, which is a ubiquitous amino acid sequence that mediates attachment of cells to many adhesive proteins.

CHAPTER 3

1. Larger proteins have more amino acids and therefore have more sites for potential interaction with a biomaterial's surface.

2. Disulfide bridges are covalent bonds that stabilize the structure of a protein. Proteins with more stabilizing cross-links are less able to unfold on a surface. Consequently, fewer binding sites would be brought into proximity to the surface to enable bonding.

3. In a single-component solution, intermolecular repulsion between the identical molecules is likely to occur. The repulsive interactions interfere with adsorption of protein molecules. In multicomponent solutions, however, attractive interactions between molecules are also possible, and increased/cooperative adsorption can occur.

4. The particular factors affecting arrival depend on the mode of mass transport, but in general important factors include concentration of the protein, its size, and flow.

5. In desorption, a protein molecule leaves the surface without influence from other protein molecules. In both displacement and exchange, competition for surface binding sites results in replacement of the protein molecule with another. With displacement, the same type of protein replaces the molecule, whereas a different type of protein replaces the molecule in exchange.

6. Large molecules, such as proteins, can form many bonds with the surface. As the number of bonds increases, simultaneous disruption of all the bonds to free the protein from the surface becomes less likely.

CHAPTER 7

1. Tissues are composed of (a) cells and (b) extracellular matrix. The type of cells and their function as well as the composition, structure and function of the extracellular matrix are specific and characteristic of each tissue.

   Examples of "hard" tissues: bone; cartrilage.

   Examples of "soft" tissues: skin; lung; vasculature; intestine; bladder; stomach.

2. This is the case of wound healing of soft tissues (under no mechanical loading).

   (a) Recruitment of neutrophils at the site of implantation involves margination of circulating neutrophils in the vasculature, adhesion to endothelial cells, diapedesis, and chemotactic migration through the tissues towards the wound/implantation site. Activation of neutrophils induses release of cytokines and/or growth factors and the process of phagocytosis.

   (b) Formation of granulation tissue: Bioactive chemical compounds secreted at the wound site during the coagulation and inflammation stages of the wound healing process induce migration, differentiation and proliferation of cells (for example, epidermal cells, endothelial cells, fibroblasts, etc.) present in the tissues on the periphery of the wound site. Cells (specific to each tissue) synthesize and deposit chemical compounds that are components (such as collagen, elastin, adhesive proteins, proteoglycans, etc.), which are characteristic of the extracellular matrix of each tissue. New blood vessels (angiogenesis or neovascularization) are formed into the repairing tissue (granulation tissue).

   (c) Possible long-term, pathological outcomes include: formation of scar tissue on the surface of skin; fibrous encapsulation of the implant; extrusion of the implant from the body; infection at the implantation site.

3. This is the case of wound healing of the soft tissue of the blood vessel wall.

   (a) Normal wound healing involves the following: blood coagulation (including activation and aggregation of platelets) and inflammation (white blood cell activation). The proliferation stage involves migration of endothelial cells from the two anastomotic sites towards the center of the vascular graft, synthesis and deposition of components of the extracellular matrix (such as collagen type III and type IV, adhesive proteins, etc.) characteristic of the vascular wall tissue. The optimal outcome is timely coverage of the inner surface of the graft with an endothelial monolayer, the only hemocompatible surface.

   (b) Complications arise due to hyperplasia (exuberant proliferation of smooth muscle cells and consequent thickening of the blood vessel wall tissue) at the two anastomotic sites. The result is stenosis of the lumen and concomitant changes in blood flow (hemodynamics). Additional complications arise due to blood clot formation at the anastomotic sites; this event leads to occlusion of the graft and cessation of blood flow.

4. (a) The desirable outcome in this case is osseointegration, resulting from formation of new bone strongly bonded onto the prosthesis. Osteoblasts, the bone forming cells, synthesize and deposit collagen type I as well as other non-collageneous

proteins (components of the organic phase of bone). Accumulation of calcium mineral on this extracellular matrix gives rise to the inorganic phase of bone.

(b) Complications arise due to the following: incomplete integration of the implant in bone; bone resorption (due to stress-shielding); loosening; particulate (wear debris) formation (inducing chronic inflammation); and infection (sustaining chronic inflammation).

CHAPTER 8

1. A drop of liquid spreading to a point at which a balance of surface tensions in the plane of the surface is reached. This indicates the tendency of a particular liquid to wet a particular surface and is related to the material's surface energy (as well as the liquid's surface energy).

2. Both involve bombardment of a surface with primary particles that results in emission of secondary particles. They differ in the type of particles. XPS involves irradiation with X rays and emission of photoelectrons, whereas SIMS involves bombardment with ions and emission of atoms, ions, and fragments of material. Both provide data about the chemical makeup of the surface, although the nature of the information is different.

3. Rather than requiring emission of secondary particles from a surface, FTIR is based on absorption of infrared radiation by bonds between atoms. Consequently, FTIR provides information specifically about the chemical bonds within the material.

4. AFM was originally developed to provide topographic information about surfaces; however, newer instruments and methods can provide information about surface chemistry and surface mechanical properties.

5. A sample is bombarded with high-energy electrons, causing emission of secondary electrons, which reflect surface topography and molecular weight of surface elements.

6. Metals tend to degrade via corrosion, polymers undergo oxidation and hydrolysis, and ceramics experience dissolution and selective leaching.

7. Morphologic—roughness, porosity
   Physicochemical—surface energy, composition, charge
   Biological—composition (e.g., peptides and proteins)

CHAPTER 9

Note: Students are encouraged to seek detailed answers (regarding, for example, specific assays/techniques and other pertinent aspects) to the questions of this chapter as Discovery Activities.

1. A haemocompatible prosthesis is a device composed of materials that meet the following criteria:

(a) minimize and/or eliminate thrombus formation (activation of the plasma coagulation cascade);
(b) minimize and/or eliminate platelet activation;
(c) minimize and/or eliminate white blood cell activation;
(d) promote endothelialization;

In addition, blood flow in the prosthesis should not cause red blood cell destruction (hemolysis).

2. (a) The objective of this endeavor is to establish the efficacy and safety of the new product. The evaluation process should include the following tests:
   - In Vitro Material Characterization (including mechanical properties, chemical composition, surface characterization, stability/corrosion, protein adsorption, etc.)
   - In Vitro Assays Using Cellular Models to test (first and foremost) for cytotoxicity (both acute and chronic) and then for various functions (including adhesion, proliferation, synthesis and release of extracellular matrix proteins characteristic of a specific cell type and tissue, etc.) of cells specific to the tissue in which the material/device will be implanted and/or with which the prosthesis will come in contact.
   - Animal Experiments to obtain in vivo information regarding inflammation (acute and chronic), the immune reaction, and carcinogenesis (long-term response).
   - Clinical Trials

     After approval by appropriate regulatory agencies (for example, the Food and Drug Administration in the United States) the device/material should be tested in human subjects.

   (b) Advantages and Limitations
   - In vitro studies are short-term, relatively inexpensive, but a very effective screening process for identifying and, consequently, removing from the study materials/devices that do not meet biocompatibility criteria. Exposure of materials/devices to one cell type at a time, however, does not simulate the competitive, complex, and dynamic physiologic milieu. As a result, the information obtained from the cellular models is valuable but limited.
   - Animal studies are complex, expensive, and difficult to interpret; they are, however, needed (as a precedent to human trials) and, thus, useful. A major responsibility of the experimenters who conduct such studies is animal welfare and humane considerations. These studies must meet institute/company as well as state/federal government regulations/laws.
   - The overall objective of the clinical trials is to establish that the new material/device is an effective and safe means to either save lives or improve the quality of life of patients without complications, pain or suffering. Since human subjects are involved, ethical issues (for example, selection of subjects from a group of patients who may benefit from the new product, confidentiality, informed patient consent, etc.) are part of the decision making process. Benefits and risks associated with the new product must be carefully evaluated. Follow-

up data must be carefully and meticulously collected, analyzed, and evaluated. The possibility of legal action taken in the aftermath of catastrophic failure of implants in humans should also be kept in mind.

(c) Special aspects:

- Cells pertinent to bone (for example, osteoblasts, osteoclasts, and fibroblasts) must be used for the cellular studies.
- Animal tests should be conducted under loading conditions (specifically, appropriately scaled prototypes should be implanted in anatomical sites in animals, and under loading conditions/use, similar to those in humans).
- Appropriate controls should be included for all experiments and for every stage of the evaluation process.
- Meticulous records of all experiments/studies should be maintained. Whenever possible, quantitative data should be collected and appropriate statistical analyses should be utilized.

3. • In this case, the biocompatibility evaluation process will include the sequence of in vitro material characterization, in vitro assays using cellular models, in vivo animal experiments, and clinical trials.
   - Polymer stability (determined, for example, by degradation experiments) should be included in the in vitro material characterization studies.
   - In vitro haemocompatibility experiments should be conducted using anticoagulated blood under flow conditions.
   - Pertinent cell types for the in vitro cellular studies in this case are endothelial cells, smooth muscle cells and fibroblasts.
   - Ex vivo animal experiments (using, for example, fistulas and/or shunts) should precede the in vivo animal studies.
   - The overall objective of these studies is described in the answer to Question #1 of this chapter.

4. (a) Functional Tests

- Require design, fabrication, and mechanical testing of completely functional animal-size versions of prostheses.
- Each material/device is placed in animals in a similar anatomical site as that of humans, and is tested in the functional mode.
- Long-term (in the order of years) studies are performed.
- *Advantages*

  Functional tests provide in vivo data under conditions that simulate those of end-use of the material/device tested.
- *Disadvantages*

  Require consideration of animal size, differences (in anatomy, hematology, histology, life expectancy, etc.) from humans.

  Require sophisticated surgical procedures, participation of experienced surgeons and veterinary pathologists, as well as of trained personnel working in the animal facilities.

Are expensive because of costs of animal purchase, long-term maintenance, and daily care.

Are complex (animal subjects are unable to actively cooperate with the experimental plan) and difficult to interpret (extrapolation of results from animal subjects to humans is inappropriate and may lead to erroneous conclusions).

(b) Non-functional Tests

- In this case, samples of arbitrary shape and size (not necessarily those of the prostheses) are tested.
- The material/device is implanted (for example, either subcutaneously or intraperitoneally) in anatomical locations in animals different than the intended end-use anatomical sites in humans. Under these circumstances, the implants are tested only in the non-functional mode.
- Short-term (in the order of months) studies are performed.
- *Advantages*

  Investigation of direct interactions between materials and the biological/chemical environment is possible.

  Access of the soft tissue sites for implantation of samples in this case requires relatively minor surgery.

  The cost of these short-duration experiments is relatively low.
- *Disadvantages*

  These tests are conducted in the absence of mechanical loading, with specimens that do not have the shape of the prosthesis.

Glossary

Abscess: Encapsulated space housing an infection within tissue.

Acquired Immunity: Resistance to pathogens developed in response to specific stimuli.

Adipose tissue: Fat tissue.

Adjuvant: Chemical that enhances the immune response to a target molecule.

Adsorption: Adhesion of a molecule to a surface.

AFM: Atomic force microscopy (or microscope).

Aggregation (of platelets): The process of platelets sticking together to form a platelet plug.

Allergen: Antigen that causes a hypersensitivity response.

Allograft: A tissue graft where the donor and recipient are of the same species; tissue taken from one person and transplanted in another person is an allograft.

Alloy: A metal containing two or more elements.

Alpha (α) helix: A common secondary structure of proteins in which the polypeptide chains is twisted into a right-handed helix stabilized by hydrogen bonds.

Amino acid: An organic molecule containing a basic amino (NH_2) group, an acidic carboxyl (COOH) group, and a side chain attached to a central α carbon atom; the building block of proteins.

Anaphylactic shock: Extremely severe physiological response to an antigen.

Anastomoses: Sites where blood vessels (or any hollow cylinders) are connected.

Anergic: Quiescent, inactive.

Angiogenesis: Formation of new blood vessels.

Antibodies: Specific classes of immunoglobulins, produced by B cells and possessing very specific and selective binding characteristics.

Antigen: Object specifically recognized by an antibody or immunoglobulin.

Antigenic determinant: The specific chemical site, on an antigen, to which an immunoglobulin binds; also known as an **epitope**.

Antigen-presenting cells: Cells (for example, macrophages) that display portions of potential antigens on their membrane in conjunction with MHC proteins, for subsequent recognition of and response to the antigen by T cells.

Apoptosis: Programmed cell death; cellular death triggered by the expression of certain genes rather than by trauma.

Arachidonic acid cascade: Intraplatelet cascade of chemical reactions that forms pro- and anticoagulants.

Arginine-glycine-aspartic acid: Also known as **RGD**. An amino acid sequence that is present in "adhesive proteins" and is recognized by cell surface receptors of the integrin family.

ATIII: Antithrombin III.

Atrophy: Reduction in size, function, or number.

Autocrine: Referring to a substance, produced by a cell, that affects the function of that cell type.

Autograft: A tissue graft where the donor and recipient are the same person; tissue taken from one site on a person and reimplanted at another site on the same person is an autograft.

Autoimmune disorders: Medical conditions in which a person develops an immune response to his/her own cells/tissues. The immune system loses the ability to discriminate between "self" and "non-self."

Autologous: Derived from the same organism.

B cells: B lymphocytes, responsible for humoral immunity.

Basophils: Nonlymphatic leukocytes, primarily important in this text for their involvement in IgE-mediated immune responses.

Beta (β) sheet: A secondary structure of proteins in which **β-strands** are aligned in a parallel or antiparallel manner, giving the appearance of a pleated sheet.

Beta (β) strand: A common secondary structure of proteins in which the polypeptide chain is in an extended conformation.

Bioactive: Able to integrate within living tissue, with direct apposition and bonding between tissue and biomaterial.

Biodegradable: Will degrade (break down, structurally and chemically) in the body from natural processes (hydrolysis, enzymatic activity, etc.).

Bioinert: Eliciting little, if any, host response and exhibiting little, if any, material changes after implantation.

Biomaterial: Synthetic or natural material that can replace or augment tissues, organs, or body functions.

–blast: Word ending denoting a primitive or formative cell; for example, a precursor cell that has the capability to differentiate into another type of cell (erythroblast) or a cell that produces a particular type of tissue (osteoblast).

B lymphocytes: Lymphatic leukocytes responsible for humoral immunity.

Cancellous bone: Bone with a porous or networked structure; also known as trabecular or spongy bone.

Cardinal signs (of inflammation): "*Rubor et tumor cum calore et dolore*," or "redness and swelling with heat and pain"; a description of the primary symptoms of inflammation.

CD: Abbreviation for "cluster of differentiation," referring to a class of cell membrane proteins.

Cell-mediated immunity: Resistance to pathogens arising from cytotoxic actions of cells.

Chemoattractant: Chemical that stimulates cell migration toward the source of the chemical.

Chemotaxis: Migration along a chemical gradient.

Coagulation: Blood clotting.

Codon: Sequence of three nucleotides in DNA or messenger RNA that codes for a specific amino acid.

Collagen: Protein rich in glycine and hydroxyproline that is a principal component of the skin, tendons, cartilage, bone, and other connective tissues.

Common pathway: Series of chemical reactions that is present in both the extrinsic and intrinsic pathways of the plasma coagulation cascade; specifically, the series of reactions from the activation of factor X down to the formation of fibrin.

Complement system: Cascade of activated enzymes that is part of the body's defense against invading microorganisms and a link between the processes of coagulation and inflammation.

Conformation: Spatial location of the atoms of a molecule; the three-dimensional shape of a macromolecule.

Corrosion: Deterioration and removal by chemical attack.

Covalent bond: Stable chemical link between two atoms produced by sharing one or more pairs of electrons.

Crenate: Scalloped or notched; a description of the puckered or wavy appearance of a cell membrane after shrinkage.

–Cyte: Word ending denoting cell.

Cytokine: Soluble molecule produced by cells that modulates the function of the cell type of origin and/or of other cell types. Originally used to refer to molecules that affected cells of the immune system (as opposed to molecules that affected the functions of other cell types), this term is now often used as a synonym for **growth factor**.

Cytotoxic T cells: T cells that express CD8 and that are able to directly attack and kill other cells.

Dalton (Da): Unit of molecular mass; approximately equal to the mass of a hydrogen atom (1.66×10^{-24} g).

Degranulation (of platelets): Release of the chemical contents of intracellular granules.

Demarcation membranes: Network of cell membrane that is dispersed throughout the volume of a megakaryocyte. When the megakaryocyte fractures, the cytoplasmic residue within the pockets of the demarcation membranes are platelets.

Denaturation: Dramatic change in the conformation of a macromolecule, such as by exposure to heat or chemicals, which usually causes loss of biological function.

Desorption: Reverse of adsorption; detachment of an adsorbed molecule from a surface.

Diapedesis: Squeezing of a blood cell through a narrow endothelial junction and into the underlying tissue, without compromising the integrity of the blood vessel.

Differentiation (of a cell): Process by which a cell undergoes a change to an overtly specialized cell type.

Diffusion: Net movement of molecules in the direction of lower concentration resulting from random thermal movement.

Disulfide bond: Covalent linkage between two sulfhydryl groups on cysteines; –S–S–.

Domain: Portion of a protein that has a tertiary structure of its own.

EDTA: Ethylenediaminetetraacetic acid—a calcium chelator.

Embolus: An object (foreign material, clot, air bubble, etc.) that moves through the circulatory system and blocks a narrow vessel.

Endothelialization: Population and coverage of a surface with endothelial cells.

Endothelium: Layer of endothelial cells that covers the interior of blood vessels.

Endotoxins: Chemicals produced in pathogens (e.g., bacteria) that are released on destruction of the pathogen and are toxic to humans.

Eosinophils: Nonlymphatic leukocytes, primarily important in this text for their parasite-destroying abilities and their ability to phagocytose antibody-antigen complexes.

Epithelialization: Population and coverage of a surface with epithelial cells.

Epitope: Specific chemical site on an antigen to which an immunoglobulin binds; also known as an antigenic determinant.

Erythro–: Word beginning denoting red

Erythrocyte: Red blood cell, responsible for oxygen and carbon dioxide transport in the blood.

ESCA: Electron spectroscopy for chemical analysis; also known as X-ray photoelectron spectroscopy (**XPS**).

Extracellular matrix: Network of proteins and polysaccharides that serves as a structural element in tissues.

Extrinsic pathway: Cascade of chemical reactions, initiated by trauma to the vascular walls and surrounding tissues, that results in coagulation.

Fatigue strength: Highest stress a material can withstand without failing during cyclic loading for a given number of cycles; also known as endurance limit.

Fibrinolysis: Enzymatic digestion/removal of blood clots via the destruction of fibrin.

Focal adhesions: Connection points between the cytoskeleton and an extracellular substrate.

Foreign body giant cells: Large, multinucleated cells formed from the fusion of macrophages; a sign of a long-term biological response to a foreign material that cannot be phagocytosed by macrophages.

Foreign body reaction: Fibrosis and the presence of foreign body giant cells surrounding an implanted object.

Foreign body tumorigenesis: Development of a tumor at the site of an implanted object because of physical rather than chemical characteristics of the object.

FTIR: Fourier transform infrared spectroscopy.

***"Functio laesa"*:** "Disturbed function," sometimes known as the fifth **cardinal sign** of inflammation, used to describe the fact that inflamed organs may dysfunction.

Genotype: Total genetic makeup of a particular cell or organism.

Glycoprotein: Protein covalently linked to at least one oligosaccharide (sugar) residue.

Granulation tissue: Highly vascularized fibrillar connective tissue, formed to fill a wound or void in a tissue as the inflammatory stage of the wound healing process draws to a close.

Granulocytes: Term referring to neutrophils, eosinophils, and basophils, based on the presence of intracellular granules in their cytoplasm.

Growth factor: Extracellular polypeptide that stimulates various activities of cells, including proliferation and/or differentiation. Often used as a synonym for **cytokine**.

Hapten: Molecule that induces little or no immune response on its own, but which can be combined with a target molecule or carrier to produce an increased immune response to the hapten/carrier complex.

Helper T cells: T cells that express CD4 and secrete lymphokines.

Hematocrit: Percentage of blood volume that is composed of cells.

Hematopoiesis: Process of blood cell formation.

Hemolysis: Destruction of erythrocytes.

Hemostasis: Regulation of blood volume.

Heterodimer: Protein complex composed of two different polypeptide chains.

Histocompatibility: Degree of similarity between the HLA of a potential tissue/organ donor and recipient, thus a prediction for the potential success of a tissue/organ transplant.

Histologic: Having to do with the small-scale structure of tissues and cells.

HLA: Human leukocyte antigen.

HMWK: High-molecular-weight kininogen.

Homodimer: Protein complex composed of two identical polypeptide chains.

Human leukocyte antigen: Another term for the **major histocompatibility complex**.

Humoral immunity: Resistance to pathogens arising from the chemical activities of antibodies.

Hydrogen bond: Noncovalent chemical bond in which a hydrogen atom serves as a bridge between two electronegative atoms.

Hydrolysis: Cleavage of a covalent bond by reaction with water.

Hydrophilic: "Water loving"; refers to polar molecules that are able to form hydrogen bonds with water and consequently dissolve readily in water.

Hydrophobic: "Water fearing"; refers to nonpolar molecules that cannot form bonds with water molecules, and therefore the molecules do not dissolve in water.

Hyperplasia: Overgrowth of cells or tissue.

Hypersensitivity: Allergy.

Hypertonic: Describing a solution with a higher than standard, reference osmolarity.

Hypertrophy: Increase in size.

Hypotonic: Describing a solution with a lower than standard, reference osmolarity.

Hypoxia: Oxygen shortage.

Ig: Immunoglobulin.

Immunogen: Object that can induce an immune response.

Immunoglobulins: Class of glycoproteins, the members of which possess regions of great structural and chemical variety, allowing very specific and selective binding of these immunoglobulins to target molecules; abbreviated "Ig."

Infection: Colonization of tissue by foreign organisms (bacteria, viruses, etc.) with some concurrent level of immune response to the organisms; symptoms of infection often also include suppuration and the signs of inflammation.

Inflammation: Term describing a normal and necessary part of the wound healing process, primarily characterized by the presence of the "cardinal signs": redness and swelling with heat and pain.

Innate immunity: Resistance to pathogens due to a variety of nonspecific physiological processes.

In situ: In the normal place or position.

Integrin: Member of a family of transmembrane proteins involved in adhesion of cells to the extracellular matrix.

Interleukins: Class of lymphokines, primarily important in this text for their ability to modulate the functions of T and B cells.

Intrinsic pathway: Cascade of chemical reactions, initiated by the exposure of blood to a foreign surface, which results in coagulation.

In vitro: In the laboratory.

In vivo: In a living being.

Ionic bond: A noncovalent bond resulting from coulombic attraction of unlike ions.

Isoelectric point: pH at which a charged molecule has no net electrical charge.

Isotonic: Solution with the same osmolarity as a standard, reference solution.

Karyo–: Word beginning referring to the nucleus.

Kinins: Class of chemicals that increase vascular permeability, stimulate pain receptors, and cause smooth muscle contraction.

Labile cells: Cells able to self-replicate.

Lamellipodia: Literally, "thin-layered feet"; used to describe broad, flat cell projections formed in the direction of cell movement.

Leukocytes: White blood cells. Note that **lymphatic leukocytes** and **non-lymphatic leukocytes** have different origins and functions.

Lumen: Inner, open space of a tube.

Lymphatic leukocytes: Leukocytes derived from lymphoid stem cells in the lymphatic system; T lymphocytes and B lymphocytes.

Lymphocytes: White blood cells of the lymphatic system; lymphatic leukocytes.

Lymphoid stem cell: Cells that migrate from the bone marrow to the lymphatic system, where they produce lymphocytes.

Lymphokines: Chemicals (such as the interleukins) that are secreted by helper T cells and regulate the functions of the cytotoxic T cells and the B cells.

Lysis: Rupture of the cell membrane that leads to cellular death.

MAC: Membrane attack complex.

Macrophages: Nonlymphatic leukocytes, mature monocytes; primarily important in this text for their functions as phagocytes and as antigen-presenting cells.

Major histocompatibility complex: Type of cell-surface protein important to the recognition of and response to antigens, which are critically important to the development of immunologic self-tolerance.

Margination: Movement of cells from the middle flow region of blood vessels to the sides of the vessels, near the walls.

Matrix metalloproteinase: Protein-digesting enzymes, in which the function of the catalytic domain of the enzyme depends on the presence of a zinc atom. Also referred to as matrix metalloprotease.

Membrane attack complex: Terminal product of the complement system, C5b6789, capable of directly lysing cell membranes.

Memory B cells: Differentiated form of activated B cells that can remain quiescent in the body for years, but which will respond quickly to the same stimulus that originally activated the precursor B cells.

Memory T cells: Differentiated form of activated cytotoxic or helper T cells that can remain quiescent in the body for years, but which will respond quickly to the same stimulus that originally activated the precursor T cells.

Metaplasia: Transformation of one cell or tissue type into another type.

MHC: Major histocompatibility complex.

Micromotion: Small amounts of slippage or movement; in this text, referring to motion at the interface between an implant and a tissue.

Mitosis: Process by which a cell splits into two new cells, each of which contains a copy of the genes of the original cell.

MMP: Matrix metalloproteinase.

Monocytes: Nonlymphatic leukocytes, immature macrophages.

Monomer: Single chemical units that are the building blocks of polymers.

Morphology: Shape.

Multipotent cell: Cell capable of producing a variety of cell types within a given category or class (i.e., capable of producing all of the different types of blood cells).

Myeloid stem cell: Cell that remains in the bone marrow and produces the precursor cells necessary to eventually yield erythrocytes, platelets, and nonlymphatic leukocytes.

Necrotic: Dead or dying.

Neotissue: New tissue.

Neovascularization: Formation of a new network of blood vessels.

Neutrophils: Nonlymphatic leukocytes; primarily important in this text for their phagocytic functions.

NK cells: Natural killer cells; cells that may be able to recognize and destroy infected cells or tumor cells.

Nonlymphatic leukocytes: Leukocytes derived from myeloid stem cells in the bone marrow; basophils, eosinophils, neutrophils, and monocytes.

Nonpolar molecule: Molecule that lacks an asymmetric accumulation of positive or negative charge and consequently is insoluble in water.

NSAIDs: Nonsteroidal anti-inflammatory drugs.

Opsonization: Binding of a molecule to an antibody on a foreign object and to receptors on a phagocyte membrane, initiating phagocytosis.

Osmolarity: Number of osmoles (molecules in 1 g molecular weight of undissociated solute) per liter of solution; in the context of this textbook, osmolarity determines the tendency of water to move across a semipermeable membrane to create equal solute concentrations on both sides of the membrane.

Osseointegration: Structural and functional connection between bone and implant in which there is no intervening fibrous tissue or chronic inflammation.

Osteoblast: Bone-forming cell that secretes the bone matrix.

Osteoclast: Large, multinucleated cell that destroys bone matrix.

Oxidation: Loss of electron density from an atom, such as occurs during addition of oxygen or removal of hydrogen.

Paracrine: Referring to a substance, produced by a cell, that affects the functions of other cell types.

Parenchymal: Referring to the specialized and essential composition of a specific tissue, rather than to general or ubiquitous components. For example, referring to skin tissue rather than to general fibrous tissue.

Passivation: Condition in which normal corrosion is impeded by a surface film.

Pathogens: Disease-causing bacteria, viruses, fungi, etc.

Peptide bond: Chemical bond formed between the carboxyl group of one amino acid and the amino group of a second amino acid.

Peptide: Sequence of amino acids; generally implying a sequence shorter than that of an entire protein.

Percent elongation: The amount of deformation before failure; also known as ductility.

Phagocyte: Cell capable of phagocytosis.

Phagocytosis: Process of "cellular eating," by which cells surround, engulf, and consume foreign particles or objects; not related to a cell's intake of nutrients.

Phagosome: Intracellular vacuole in which phagocytosed objects are degraded.

Phenotype: Observable characteristics typical to a given cell or organism type due to the selective expression of genes (for example, the production of certain amounts/types of proteins by cells).

PHSC: Abbreviation for "pluripotent hematopoietic stem cell."

Plasma cells: Differentiated B cells that secrete antibodies.

Plasma: Noncellular portion of blood; a solution of salts, proteins, gases, and other biomolecules in water.

Platelet plug: A mass of aggregated platelets that "plugs" a defect in the endothelium or blood vessel or that covers a foreign surface.

Platelet: Blood-borne cellular fragment that retains enough biochemical activity to adhere to an injured blood vessel or a foreign surface and release chemicals that cause other platelets to aggregate at that site.

Pluripotent cell: Capable of replicating and differentiating into multiple types of cells.

Polymer: Molecule composed of repeating chemical subunits.

Polymorphonuclear: Having a lobed nucleus or nuclei.

Polypeptide: Linear polymer composed of multiple amino acids.

Primary structure: Linear sequence of amino acids in a protein.

Proliferation: Cell division or replication to form more cells.

Protease: Enzyme that breaks down a protein or proteins.

Protein: Large molecule composed of one or more chains of amino acids.

Proteoglycan: Molecule containing peptide and lipid domains.

Proteolytic: Protein degrading.

Pseudopodia: Literally, "false feet"; used to describe limblike cell projections; encompasses filopodia and lamellipodia.

Pyrogens: Fever-causing agents.

Quaternary structure: Three-dimensional relationship of two or more polypeptide chains in a protein complex.

Residue: General term for the unit building block of a polymer; with respect to proteins, this refers to an amino acid.

Resorption: Destruction of bone matrix.

Reticulo–: Word beginning referring to the property of having a network of filaments.

RGD: Amino acid sequence arginine-glycine-aspartic acid, which is present in "adhesive proteins" and is recognized by cell surface receptors of the integrin family.

Rouleaux: "Stacked-up" aggregates of red blood cells.

"*Rubor et tumor cum calore et dolore*": "Redness and swelling with heat and pain," or the cardinal signs of inflammation; a description of the primary symptoms of inflammation.

Secondary structure: Regular local folding pattern of a protein, such as an α-helix or β-sheet.

SEM: Scanning electron microscopy.

Sensitization: First exposure to an allergen.

Shear rate: Speed at which shear strain is applied; how quickly a given amount of shear deformation is induced.

SIMS: Secondary ion mass spectroscopy.

Sinus: Empty space or cavity.

Static cells: Cells incapable of self-replication.

Stem cell: Cell capable of replicating itself (while staying the same cell type) as well as differentiating into other cell types.

Stent: Annular framework placed inside a blood vessel to hold the walls of the vessel open and allow blood flow.

Stroma: Three-dimensional network of spongy tissue in the bone marrow, made up of cells and structural fibers.

Substratum: Underlying material; a base or substrate.

Suppuration: Pus production.

T Cells: T lymphocytes; responsible for cell-mediated immunity.

Tertiary structure: Complex three-dimensional structure of a protein governed by interactions between regions more distantly separated on the polypeptide chain.

Thromboembolism: Clot that moves through the circulatory system and blocks a narrow vessel.

Thrombus: Clot.

Tissue typing: Determining potential matches between organ/tissue donors and recipients.

T lymphocytes: Lymphatic leukocytes responsible for cell-mediated immunity.

Totipotent cell: Capable of producing any other cell type.

Ultimate compressive strength (σ_{UCS}): Maximum load per unit area accommodated by a material loaded in compression.

Ultimate tensile strength (σ_{UTS}): Maximum load per unit area accommodated by a material loaded in tension.

Vascular spasm: Constriction of blood vessels, due to injury, that stops or slows blood flow.

Wear debris: Small pieces of material generated by two articulating surfaces (for example, a femoral head and an acetabular cup) grinding together over time.

Xenograft: Tissue graft where the donor and recipient are not of the same species; porcine tissue transplanted into a human being is a xenograft.

XPS: X-ray photoelectron spectroscopy; also known as electron spectroscopy for chemical analysis (ESCA).

Young's modulus: Modulus of elasticity in tension; stress per unit strain; the slope of the initial linear (elastic) portion of a stress-strain curve.

Index